Enterprise Information Exchange

A Roadmap for Electronic Data Interchange for the Manufacturing Company

V.O. Muglia, CSI
Editor

Published by the

Computer and Automated Systems Association
of the Society of Manufacturing Engineers
Publications Development Department
Reference Publications Division
One SME Drive
P.O. Box 930
Dearborn, Michigan 48121

Enterprise Information Exchange

A Roadmap for Electronic Data Interchange for the Manufacturing Company

Dearborn, Michigan 48121

First Edition
First Printing

Library of Congress Catalog Card Number: 92-085523
International Standard Book Number: 0-87263-435-3
Manufactured in the United States of America

about CASA/SME

The Computer and Automated Systems Association of the Society of Manufacturing Engineers (CASA/SME) was founded in 1975 to provide comprehensive and integrated coverage of the field of computers and automation for the advancement of manufacturing.

As an educational and scientific association, CASA/SME has become ''home'' for engineers, managers and other professionals involved in computer-based technologies and automated systems. CASA/SME is applications-oriented and addresses all phases of research, design, installation, operation and maintenance of the total manufacturing enterprise. This book is one example of its wide-ranging activities.

Specific CASA/SME goals are: (1) provide professionals with a focus for the many aspects of manufacturing which utilize computer systems automation; (2) provide liaison among industry, government, and education in identifying areas of further technology development; and (3) encourage the development of the totally integrated manufacturing enterprise.

TABLE OF CONTENTS

Page

Introduction 1

1. Futures and Strategic Positioning

21st Century Business Information Outlook 5
by V.O. Muglia, CSI
Caterpillar Company, Solar Turbines Incorporated Subsidiary

Information Systems Strategy: A Case Study within a Mature Industry 11
by Frances A. O'Connor, Richard Diesslin, and John Lamoureux
Science Applications International Corporation (SAIC)

2. EIX and Integration Basics

EIX Definitions and Framework 33
by Howard E. Groesbeck
Groesbeck and Associates

A Roadmap for Enterprise Integration 37
by Dr. Richard J. Mayer and Captain Michael K. Painter
Texas A & M University, United States Air Force

3. Industry and Government Standards

Implementing Standards for Interdepartmental Document Sharing 57
by Dano Ybarra
OMS, Incorporated

An Overview of Electronic Data Interchange Standards 61
by Leslie Rohde
The LJR Group

PDES: The Enterprise Data Standard 67
by Robert A. Carringer, CMfgE
Institute of Business Technology

Computer-Aided Acquisition and Logistics Support (CALS) Primer 75
by Paul N. Pechersky
E-Systems Incorporated

4. Planning Information Exchange within the Organization

Engineering Data Management 85
by Sharad Sheth
Electronic Data Systems Corporation

Page

A Flexible Manufacturing Technical Data Management System 93
by Mark J. Beach and Alan C. Jones
IBM Corporation

Automating Electronics Manufacturing Documentation Management 103
by Gerald Ginsberg, PE
Component Data Associates Inc.

5. Planning Information Exchange External to the Organization

Electronic Interchange of Product Definition Data Between Companies 111
by Robert A. Carringer, CMfgE
International TechneGroup Incorporated

EDI From a Supplier's Viewpoint 119
by Lee K. Foote
E.I. Du Pont de Nemours and Company

The Success of Customer/Supplier Information Exchange 123
by Mary K. Johnston
IVAC Corporation, Eli Lilly & Company Subsidiary

6. Information Technology Enablers and Systems Planning

Information System Architecture and Enablers for Enterprise Information Exchange 129
by M.J. Quint
Perceptive Solutions Incorporated

Integration of Product Information Residing on Various Computer Systems 133
by Atul C. Patel
Systech, Incorporated

Distributed Databases in a Heterogeneous Computing Environment 147
by Lawrence A. Rowe
University of California at Berkeley

Which Network is the Right One? 155
by James G. Ames
Arizona State University

7. Justification

World Class Cost Management and CIM Justification: The Way of the 90s 163
by David O. Nellemann
Andersen Consulting

8. Reference Bibliography 175

Index 185

INTRODUCTION

Sharing and exchanging information is core to a business providing a product or service and is accomplished either by design or default. Often the quality of the information and its exchange corresponds with the enterprise's success or failure. Since there is no equation or formula for a successful system, it is important to begin a system design with a quality planning process and, as extensive as possible, relevant references. This book is intended to provide a set of quality references to enable the manufacturing engineer, manager, information technology system specialist, or design team participant to enhance her/his EIX (Enterprise Information eXchange) communicating, planning, alternative analysis, and decision-making activities.

This reference book is organized into eight sections:

- Section One: *Futures and Strategic Positioning.*
- Section Two: *EIX and Integration Basics.*
- Section Three: *Industry and Government Standards.*
- Section Four: *Planning Information Exchange within the Organization.*
- Section Five: *Planning Information Exchange external to the Organization.*
- Section Six: *Information Technology Enablers and Systems Planning.*
- Section Seven: *Justification.*
- Section Eight: *Reference Bibliography.*

This topic requires several texts to serve as a detailed instructional guide. Consequently, each section contains articles providing a significant component of an individual's EIX knowledge base. Additional recommended references for further understanding are identified in the Reference Bibliography.

Many people influenced and/or contributed to the implementation of this book. The editor wishes to thank the members of the Computer and Automated Systems Association/Society of Manufacturing Engineers (CASA/SME) Board of Advisors for their insight as to the need for this document; Mr. Robert King and staff of SME's Publications Development Department for their support during the preparation and delivery of the final materials to the printer; the authors for their knowledge and motivation to share; and Ms. Nancy Mauter, of SME staff, for her headquarters liaison, writing encouragement and positive attitude, without which this book would not exist.

Victor O. Muglia, CSI, Editor
Caterpillar Company
Solar Turbines Incorporated Subsidiary
San Diego, California
September 1992

Section One:
Futures and Strategic Positioning

Section One papers are:

- *21st Century Business Information Outlook.*
- *Information Systems Strategy: a Case Study within a Mature Industry.*

To begin a basic understanding of an EIX environment, the future environment should be considered before planning any change. Muglia describes the significant 21st Century information environment as a reference for planning information systems visions and directions. The future emphasis on information creates a significant requirement for effective information transfer in the next century. His predictions of new improved systems and services incorporate challenges that planners and managers need to consider if future information systems are to be successful.

Considering the future possibilities and requirements, a business information strategy is needed. A case study involving Navistar's Information Systems Strategy, as a mature industry, is presented by O'Connor, Diesslin, and Lamoureux. A mature industry strategy is one of the hardest strategies to develop and serves as a quality reference. This paper describes the study process and information strategy framework needed by a manufacturing enterprise. The content will assist in establishing strategies for implementing new and enhanced legacy systems.

21st Century Business Information Outlook

V.O. MUGLIA, CSI
Caterpillar Company, Solar Turbines Incorporated Subsidiary

"Knowledge is Power," is an often-heard phrase. The ability to have relevant knowledge and information when, where, and in the desired format enables workers to meet any decision or analysis needed. Dramatic improvements in information technology price performance is increasing into the 21st Century. Computer processing speed today is expected to increase by 100 times by the year 2000. A two-inch cube will be able to store the contents of 100,000 books. The ability to capture, retain, and reuse knowledge of experts will be commonplace.

Benjamin and Blunt envision the following scenario:

"It's a Monday morning in the year 2000. Executive Joanne Smith gets in her car and voice activates her remote telecommunications access workstation. She requests all voice and mail messages, open and pending, as well as her schedule for the day. Her workstation consolidates the items from home and office databases, and her "message ordering knowbot," a program she has instructed, delivers the accumulated messages in the order she prefers. By the time Joanne gets to the office she has sent the necessary messages, revised her day's schedule, and completed a to-do list for the week, all of which have been filed in her "virtual database" by her "personal organizer knowbot." (1)

James Emory states, in MIS Quarterly (4), that this situation can be considered a "magic genie" for information accessibility. We currently see the rudiments of this "magic" in global systems, such as telephone systems, corporate EDI, and satellite information transfer systems and in the rapid change in our commonplace personal computers, information kiosks, and intelligent products.

Alvin Toffler and Shoshana Zuboff talk of the next century as an "Information Age." Toffler states that:

"Knowledge is change...and accelerating knowledge, fueling the great engine of technology, means accelerating change. " (9)

John Zachman of IBM proclaimed that:

"Soon, the enterprise of the information age will find itself immobilized if it does not have the ability to tap the information resources within and without its boundaries. " (11)

What does this mean to the 21st Century business? This paper will provide a vision for the future based on technology and culture trends, and list some challenges to business, as well as to government, education, and the community.

FIRST, A FEW PREDICTIONS FROM LEADING EXPERTS

Let's look at the predictions of several individuals before exploring the impact on business and the rest of the world.

Cunningham and Porter discuss the role of ISDN, the Integrated Services Digital Network, a two-way cable service. They believe this will be an established part of society by the year 2000. The ramifications they see are:

1. Greater user freedom to select TV, CDs, computer bulletin boards, fax, electronic mail, databases, newspapers.

2. Interconnection of diverse databases to expand information to many more people.

3. The need to establish protection systems for data, information, knowledge, networks, and owners.

4. The need to address issues related to access restrictions due to pricing, and information manipulation of the "haves" versus the "have-nots."

5. Information transfer worldwide can change current relationships with corporations, governments, cultural, and religious institutions.

6. Information and knowledge accessibility can speed political, economic, cultural, and corporate change.

7. Electronic transfer of money could lead to stability issues.

8. Telecommuting could keep workers at home.

9. Social lives could be enhanced with ease of communication or degraded if interactions are only via networking.

10. Care must be taken to avoid the trend toward people-as-numbers. (3)

Lundberg discusses the future preparations of Inmarsat, the global mobile satellite communications provider. They have included the following in their plans for availability by the year 2011:

1. A global personal communicator, a pocket-sized telephone.

2. Global satellite paging.

3. Voice recognition customized personal computers.

4. Picturephone.

5. Databases, Information Bases, and Knowledge Bases that replace libraries and can be accessed throughout the world.

6. A worldwide network of environmental management.

7. Long distance charges will be abolished by December 31, 2000 and replaced by all local calls due to this mobile, global environment.

8. ISDN (Integrated Services Digital Network) integrated with mobile satellite features. (7)

Benjamin and Blunt look at a broad perspective including technology, architecture and standards, services, economics, applications, and change management. Their predictions for the next decade are:

For Technology: The cost performance of everything related to Information Technology (e.g. memories, microprocessors, etc.) will improve by two orders of magnitude. In addition, the billion bit backbone network will be completed; it will be the international highway of business communication.

For Architecture and Standards: Client/server will be the predominant technology architecture, and it will evolve into an important application architecture.

For Services: Electronic mail will become ubiquitous, integrating graphics, voice, and text, and it will provide extensive collaborative support capabilities.

For Economics: Major investments will be made to complete and maintain the infrastructure. And because technology is increasingly cheaper for all, the advantage will go to those who (a) apply it well and (b) effectively purchase value-added services for implementing it.

For Applications: Applications will be designed and built using high-level business models. Emphasis will be on design of robust applications that adapt to both short-term operational difficulties and evolutionary change. In addition, the implementation process within and between large businesses is generating larger and more complex applications. Because the design issues are so complex, it is reasonable to expect one or two application Chernobyls.

For Change Management: The executives in charge of Information Technology organizations will have to learn change management skills and make sure that these skills are built into the Information Technology organization. (1)

Halal describes the "virtual community" as "a web of social/ information networks spanning the globe." (5) Like Cunningham and Porter, he sees a shift of power to those who have information and can use it. He's provided a Delphi Forecast for Information Technology, which is included in Figure 1.

Michael Scott Morton's research at MIT's Sloan

Milestone	Year
Sophisticated software programs are developed for personalized teaching, managing medical care, total control of corporate operations, etc.	1996
Expert systems are commonly used to make routing decisions in business, engineering medical diagnosis, and other fields.	1998
Access to library materials via computer is more convenient and less expensive than going to the library.	2000
Optical computers enter the commercial market.	2000
Small computers about the size of a writing pad are commonly used by most people to manage their personal affairs and work.	2002
Voice-access computers permit faster, more convenient interaction between humans and machines.	2002
Education is commonly conducted using computerized teaching programs and interactive TV.	2002
Public networks permit anyone access to libraries of data, information, electronic messages, video teleconferencing, common software programs, etc.	2003
Routine parts of most software are generated automatically.	2003
Parallel processing using multiple chips becomes dominant.	2003
Computer programs have the capacity to learn by trial and error in order to adjust their behavior.	2004
Teleconferencing replaces the majority of business travel.	2006
Half of all goods in the United States are sold through computer services such as Prodigy.	2007
Half of all U.S. workers perform their jobs partially at home using computer systems.	2009

Figure 1. *Halal's Delphi Forecast for Information Technology. (5)*

School of Management in cooperation with 10 leading corporations derived the following six findings as having a major impact on leading businesses in the mid-1990s to early 21st Century.

1. Information Technology is enabling Fundamental Changes in the Way Work is done.

2. Information Technology is Enabling the Integration of Business Functions at All Levels within and between Organizations.

3. Information Technology is Causing Shifts in the Competitive Climate in Many Industries.

4. Information Technology Presents New Strategic Opportunities for Organizations That Reassess Their Missions and Operations.

5. Successful Application of Information Technology Will Require Changes in Management and Organizational Structure.

6. A Major Challenge for Management in the 1990s Will Be to Lead Their Organizations through the Transformation Necessary to Prosper in the Globally Competitive Environment. (8)

BUSINESS IMPLICATIONS

We have seen what various individuals have predicted for our future. Now let's look at some business uses and ramifications.

Accessibility

The following six points should be made in the area of Accessibility.

1. Decision-making can be brought down to lower levels in organizations. This will empower workers, making them more satisfied business partners and also more valuable to the enterprise. In turn, this development will free management to plan, direct and lead the organization into the future. Their time fighting fires and making routine decisions will diminish. Therefore, their time can be used more constructively and innovatively to address competitive pressures, future market needs, and listening to the voice of the customers.
2. Expert systems designed to bring the best information about particular topics together, can provide the worker with high quality information and knowledge needed to make decisions and solve problems.
3. Books, databases, up-to-date information, and expert knowledge on any topic will encourage a learning environment. (6)
4. Rural areas will have increased access to the same information as urban areas in a "Rural Area Network (RAN)." (10) Providing the same educational opportunities and business data here will enable these areas to compete successfully for resources, including personnel. Increased two-way communications could also have political implications.
5. Providing for the development of cultural minorities through enhanced communications in an holistic perspective could change the way we currently interact. (2)
6. Education can be personalized to meet the needs and abilities of the individual. Access to educational opportunities will not be dependent on location.

Social/Societal

The following social-related points should be made.

1. Global communications could increase travel due to the ease of accessibility. However, decreased travel may result for business purposes due to the increased information availability, teleconferencing, and picture phones.
2. Telecommuting will decrease traffic, the need for corporate office space and parking. This trend will require changes in current management policies and practices to track work completion and provide feedback.
3. Empowerment of the individual worker will change current hierarchies within organizations—service, technical, educational, governmental—to a partnership with owners/managers/workers, administration/instructors/students, and governments/governed.
4. The use of networks for shopping, voting, education, and work will decrease the human interactions currently in these activities. Customer service needs will greatly increase as a result. With the probable ease of switching vendors, consumer loyalty will have to be cultivated.
5. With less commuting time, "the home may recover its traditional role as a center of production," per Halal. (5)
6. Halal also sees a "resurgence of cohesiveness in family life, neighborhoods, and cities." (5) As increased time is spent in the home, the individual's focus will be directed toward it and the surrounding community.
7. Communication worldwide will enhance understanding of the diversity of peoples, enabling the best use of many talents.

Organizational Changes

Businesses, as well as governments and educational institutions, will need to change to meet the changing needs of their people.

1. Service to employees and customers will become a top priority to enable productivity, growth, and loyalty.
2. Management focus must move from control to leadership.
3. Empowerment of workers/students also must include teaching responsibility and accountability. Reward systems must meet the needs of the reward/award recipient.
4. Increased connectivity and cooperative information systems will enable improved interorganizational relationships and enhanced group productivity.

5. Continual education/skill development will be necessary for all workers, enabling the ability of the organization to grow and change as the environment dictates.

6. Educational programs must partner with government, industry, and service organizations to provide needed prerequisites and continuing skill development. In this way, monetary support to educational institutions would provide individuals with skills needed by the organizations supporting them.

7. Students would contract with business or government to provide educational support in return for work within the organization.

8. Management - worker relationships must be developed via networks rather than a majority of face-to-face contacts. The staff meeting may be conducted via teleconferencing.

9. Work that can be done off-site could be accomplished anywhere in the world. Global workforce could be a reality.

10. Suppliers, manufacturers, and customers will collaborate to improve their product's total value chain leveraging well-engineered integrated information and knowledge systems.

CHALLENGES

As these multitude of changes take place, there are many challenges businesses will face. Listed below are the more significant ones.

1. The most important challenge is preparing our organizations and ourselves for the speed of progress in the next decade. Business workers must become experts at managing and accepting change.

2. Next, businesses must strategize, engineer, and build an information and knowledge infrastructure on which all employees will be able to easily access, contribute, and use.

3. Successful enterprises will create a business environment where managers lead and workers continuously learn, collaborate together, adapt, and improve leveraging the thought processes, information, knowledge, and insights of the best experts.

4. To support the knowledgeable worker, leading companies will create a cost effective integrated Information Technology architecture and infrastructure that a) supports the business's short and long term needs and b) enables easy transparent access to and use of all applications and bases internally or externally when and where needed.

5. For increased effectiveness, an organization should seamlessly integrate Human Resource's and Information Technology's cognitive, managerial, and behavioral functions into a cohesive new force ahead of their competition.

If the business focus is not directed toward these enhancements, their potential will be greatly reduced and the competitive gap will widen. The current business "haves" will far outdistance the "have-nots," requiring extensive remedial work to reduce the "one-sidedness" of the business users taking advantage of the power of knowledge and information.

REFERENCES

1. Benjamin, Robert I., and Jon Blunt. "Critical IT Issues: The Next Ten Years." *Sloan Management Review.* Cambridge, MA: Sloan School, MIT. Summer 1992.

2. Casmir, Fred L. *Communication in Development.* Norwood, NJ: Ablex Publishing Co. 1991.

3. Cunningham, Scott and Alan L. Porter. "Communication Networks: A Dozen Ways They'll Change Our Lives." *The Futurist* 26:1. Jan-Feb 1992, 19-22.

4. Emory, J.C. "Editor's Comments." *MIS Quarterly.* December 1991, xxi-xxiii.

5. Halal, William E. "The Information Technology Revolution." *The Futurist.* July-Aug 1992, pp. 10-15.

6. Horner, Vivian M. and Linda G. Roberts, eds. "Electronic Links for Learning." *The ANNALS of The American Academy of Political and Social Sciences.* vol 514. March 1991, 1-174.

7. Lundberg, O. "The Perils of Being a Visionary: One Man's Vision." *Intermedia.* 19:1. Jan-Feb 1991, 33-39.

8. Scott Morton, Michael S. *The Corporation of the 1990s: Information Technology and Organizational Transformation*. New York: Oxford University Press, 1991.

9. Toffler, Alvin. *Future Shock*. Random House, New York, 1970.

10. U.S. Congress, Office of Technology Assessment."Rural America at the Crossroads: Netwroking for the Future". Washington: USGPO. May 1991. (S/N 052-003-01228-6).

11. Zachman, John A. and John F. Sowa, "Extending and Formalizing the Frame work for Information Systems Architecture." *IBM Systems Journal*, Vol. 31, No. 3, 1992.

BIBLIOGRAPHY

Benjamin, Robert I., and Jon Blunt. *The Information Technology Function in the Year 2000: A Descriptive Vision*. Cambridge, MA: Center for Information Systems Research, MIT. March 1992.

National Academy of Engineering. *People and Technology in the Workplace*. Washington: DC. National Academy Press, 1991.

Scientific American. "Communications, Computers and Networks"." *Scientific American*: September 1991. This issue is devoted to a series of articles on how computers and telecommunications are changing the way we live and work.

Zuboff, S. *In the Age of the Smart Machine: The Future of Work and Power*. New York: Basic Books, 1988.

Presented at the CASA/SME AUTOFACT '90 Conference, November 12-15, 1990

Information Systems Strategy: A Case Study Within a Mature Industry

FRANCES A. O'CONNOR, RICHARD DIESSLIN, JOHN LAMOUREUX
Science Applications International Corporation (SAIC)

A major information systems modernization effort is described. The effort is aimed at improving the delivery of information to the shop floor. This study concentrated on the Navistar Springfield Assembly Plant and the problem of in-plant information systems and support of the systems. The study was performed over a six-month period with a small team of two Navistar personnel and two consultants from SAIC's Solion Division. The analysis used USAF ICAM Definition Language functional (IDEFO) and information (IDEFI) modeling as the analysis methodologies to establish working models for consistency and communication. The study began with an evaluation of existing systems which supported the plant internally and externally. Specific reports, forms and data screen were studied and categorized. Key personnel in the shop were interviewed. Procedures, or a perceived lack of procedures, were of higher concern to individuals than the need for new systems. This overview of the information systems study provides an outline on developing an information systems strategy based on manufacturing needs.

INTRODUCTION

The Distributed Systems Architecture Study was conducted by the Solion Division of Science Applications International Corporation (SAIC) for Navistar International Transportation Corporation, the North American market leader in medium and heavy duty trucks. The objective was to develop a Roadmap for the modernization of the information systems environment at Navistar's largest assembly plant, located in Springfield, Ohio. Information systems are important because effective use of product and process information is a major competitive weapon, especially in a mature industry. The company which most effectively utilizes its information will likely dominate the market.

The Roadmap provides a common vision so that decisions can be made within the context of an overall plan. Navistar is not unique in the fact that MIS, Engineering, and Manufacturing often have different perspectives, but a truly integrated enterprise requires that these be aligned toward a common goal. If progress toward this is to be made, a plan is needed to ensure that each piece fits into the future. Also, the complexity of the existing systems environment cannot be overemphasized a change in any of Navistar's major corporate information systems significantly impacts the Springfield Assembly Plant and vice versa. What is needed is a step-by-step migration plan which provides a means of change impact analysis and a yardstick by which to measure progress.

The approach taken was to perform a top-down analysis, resulting in a bottom-up implementation plan. The intent was not to present a multi-year, multi-million dollar solution—it was to provide the beginnings of an overall plan under which many small improvements, as well as large improvements, can be coordinated. Payback is then immediate. The methodology used was to analyze the existing environment, conceptualize a future environment, perform a needs analysis (based on the difference between the two), develop a set of recommendations, and propose a phased project plan to move forward. The view taken was that information flow is a process with inputs, outputs, and feedback mechanisms.

The complexity of the existing systems environment, the sources and impact of change, and the need for immediate improvements led to several conclusions. First, the plan must be evolutionary, provide for orderly migration away from obsolete systems, and be adaptable to constant change. Second, future systems development must focus on effectiveness (functionality and flexibility) over efficiency (minimum CPU time and disk space). Third, a primary thrust must be to decouple applications and information. Currently systems include both, and the future objective is to have information stored on the network for all to use. Finally, basic information needs will remain relatively constant over time. Most improvements are made in the process, not in the inputs and outputs. Because of this, incremental improvements can be made by making existing information available in more flexible ways.

A major area for change was in the plant culture. Automated systems will not improve poor procedures and simply executing the proposed project plan will not warrant the investment. The emphasis must be on simplification of existing business and information flow processes. Discipline will be required. On-line integrated information will require that all persons responsible for inputting data do so accurately and faithfully, or widespread failure can occur. If the culture cannot be changed, then no plan will succeed.

The developed Roadmap took a first cut at the pieces needed for migration: a Functional Organization (how the plant is set up), the Activities (what the plant does), the Information Flow (what the plant needs to know to do it), an Applications Portfolio (systems to support the information flow), a Conceptual Data Model (the database schema), a Hardware Architecture, and a Migration Plan (for the existing systems).

Four major recommendations were proposed. First, an open systems architecture should be followed. While the technology is relatively new, it is a strategic direction to allow for use of the most appropriate vendor for any given problem. Second, a continuous education plan for the plant needs to be developed which will teach systems discipline, and more importantly, promote a culture conducive to change. Third, this Roadmap should be expanded and kept updated on an ongoing basis (in an automated tool) and used to communicate between various disciplines and functional organizations. Finally, the overall resource commitment to achieve the vision of the future should be established. Based on Corporate financial commitment, the timing of the migration and the expected progress can be communicated.

If this Roadmap is kept current and used as a vehicle for communication, it can promote a common vision and make explicit those issues which are often the source of great confusion and frustration. This will enable Navistar to begin to truly integrate business, engineering, and manufacturing information in an atmosphere which encourages innovation and change in the information systems environment.

OBJECTIVES

The purpose of this paper is to document an approach to the implementation of flexible information management systems for Navistar's Springfield Assembly Plant. Critical to the overall success of this effort was the willingness of Navistar management to accept the need for changes in the environment as a whole with regard to policy, procedures, and information exchange. We drew a picture of the current environment and a vision of a possible future environment. The projects defined were listed as a way to move from here ("As Is") to there ("To Be"), but the projects alone are not the deciding factor in achieving success.

The path to integrated systems is through integrated management and streamlined functionality. The Springfield Assembly Plant is an island dependent upon corporate support functions in order to perform many tasks. Communication between plant and corporate support functions is as critical as communication between internal plant support functions and actual production departments. Information was treated in a piecemeal fashion, without an overall plan. Product and process data are two of the most important assets that Navistar owns. They are the competitive edge for any company in a mature market. Inefficient utilization of the data can be a major source of low productivity, low quality, high direct cost, and high overhead. For example, consider the tremendous impact of inaccuracies in product design and specification when exploded into material requirements, and the inability to get information when and where it is needed to support actual production/assembly steps. The future environment must support integration within each special process area (paint, axle trim, etc.) and among them as well. The traditional general business management processes, such as marketing, finance, and human resources, also have to be actively included in plant operations. Everyone in every area must have the information resources needed to do the job.

If any company is to prosper in the future, it will be because it develops an approach to plan for and stimulate change. The management must be able to address issues of cost, quality, and schedule from the top down. Such issues are timeless—the ability to produce what the customer wants on time, with good quality, and make a profit. The difference between

"As Is" and "To Be" comes in developing tools to support management in addressing these issues. Information management is one of the most important tools.

The goal of the Distributed Systems Architecture Study was to produce a Flexible Information Systems Architecture (FISA) which provides flexible information as needed (e.g., just in time) which is necessary, sufficient, and accurate for truck assembly and plant management. This needed to be done within the context of an overall plan—A Flexible Information Systems Architecture (FISA). This evolutionary plan had these five primary objectives:

- Provide for a means of integrating legacy (existing) systems within the factory.
- Provide a methodology for developing new systems which are integrated with the legacy systems.
- Provide maximum flexibility to the end user.
- Show visible (measurable) benefits in terms of cost, quality, and schedule.
- Remain consistent with both the Navistar corporate Information Systems Strategy and the Manufacturing Systems Strategy.

The focus of the Distributed Systems Architecture Study was on the Springfield truck assembly facility and its business units. This included the requirements to support the corporate information systems and to be compatible with other modernization efforts.

General Environment

FISA technology as a support system cannot start with a clean slate. Many interfacing and integrating factors complicate the scope and the succinct boundary definition. FISA is not an application system but must provide for the demand of various users and developers. Outside factors will influence FISA development and users. For example, the longevity of computer technology currently is only three to five years between major changes. On the other hand, once built, computer applications have a tendency to last as long as 15 to 20 years. Due to the dichotomy between changing computer technology and user specific requirements, companies are continually modifying their computer programs. Just as many refinements have occurred in computer technology, many improvements have been made in the production of computer hardware to drive the costs down. The corporate expenditure is still increasing due to capacity requirements as evidenced by the growing pervasiveness of computing workstations. An important consideration for FISA is the projected computer literacy factor which is rapidly increasing. The increase of computer literacy and the availability of very sophisticated and relatively inexpensive applications packages will bring about the demand for automated intelligent integration access and analysis tools. These and other factors known by the vendors, users, and integrators, along with those discovered during the requirements analysis, will clearly and succinctly define FISA and the challenging issues for success.

Huge conglomerates with very complex natures compete for market share in the business space. These businesses are geographically dispersed, cohabitating, and multi-disciplined competitors who establish partnerships and teaming agreements under changing management styles to remain leaders in the national and international markets. The demands for shared, accurate, and timely decision making information at all levels of the organization are intense. These demands are causing changes in systems applications and instrumentation in many segments of businesses. In addition, these demands are requiring the businesses to infuse new technologies into their structures. The islands of new technology are in turn creating islands of information. One of the reasons for developing integratable systems and integration tools is the lack of organic talent in the business base and, perhaps more importantly, the lack of integration expertise and integrated implementation tools. The environment in which future systems will be built is complex, intense, and typical of the large business profile.

Specific Environment

Many large, progressive companies such as Navistar provided centralized mainframe computer systems and staff to support information systems in the 1970s and early 1980s. This was the most cost-effective and staff-effective approach for that time frame. Distributed processing is now becoming more and more cost-effective and is opening up whole new approaches to communications and management information systems. It is now desirable for Navistar to pursue a corporate strategy which promotes distributed data processing, providing a multi-vendor platform/compatibility and flexible systems utilization.

A quote from the Air Force sponsored project,

Integrated Information Support System, summarized accurately much of Navistar's needs:

"Information - A Wasted Resource: Today's factories are characterized by a multiplicity of discrete information systems which have been designed to serve individual users or individual activities within an organization. In today's environment, each user receives information, massages it, generates his own information and passes it to another activity. The second activity combines it with other information, and passes it along. Since many of these individual activities receive information from and generate information for many other activities, an extremely complex information system has been developed. This complexity is compounded by the fact that information is inaccurate, untimely, and unshared.

Additionally, today's environment is characterized by a number of different computers which do not easily communicate, which have different keyboard access, and which create careers for experts on specific computers. This expertise leads to a reluctance to tackle other types of equipment.

While capital spending for computer equipment and software is soaring and information costs are mounting rapidly, the current environment is resulting in the poor treatment of data and waste of information, a valuable corporate resource. This situation is worsening as more discrete systems are put into place."

The average age of Navistar corporate computer systems was eight years, with a range extending up to 15 to 20 years. Many of these applications need to be redesigned over the next five to 10 years. The key is to accomplish data integration by designing new applications, which can be integrated with existing applications. Current methodologies and technologies are becoming available which enable this kind of integration to evolve in a distributed environment over different types of hardware. The Flexible Information Systems Architecture plan begins now and builds in a compatible manner toward this kind of future plant/corporate-wide distributed, integrated information systems environment.

Overview

Figure 1 represents the overall approach taken to define the Flexible Information Systems Architecture for the assembly plant. The scope had been defined as the four walls of the Springfield Assembly Plant and its corporate information support.

The approach may be viewed as a "look back" at the past systems and organization structure with a "vision" for the ideal future environment. This documented study itself represented a first cut at an integrated information strategy, setting goals and conceptualizing projects critical to design and implementation for the next five to 10 years.

The tasks undertaken to define the FISA were: defining the current environment, conceptualizing the future environment, analyzing the current needs to achieve the future goals, recommending solutions, and actually developing the plan itself (see Figure 2)

In the current and future environments, the manufacturing functions were studied in terms of the activities performed within each function. The information used to support the activities was defined and then the systems which create/process/support that particular information were identified. In both environments, it is important to understand all three aspects: activity, information, and actual systems. In this way, one can understand the functions and information requirements for the plant and then figure out how to satisfy these requirements.

The Recommendations section lists the alternative technologies and methodologies that were considered and then presents an evaluation of each. The most applicable solutions are included as recommendations.

The Implementation Plan section outlines a step by step approach to implementing the technologies and process changes which will lead Navistar from the current environment to the future environment. The plan outlines the projects in terms of priorities and potential benefit. Once the plan is adopted, it must be kept up to date as systems are changed, created, and eliminated.

CURRENT ("AS IS") ENVIRONMENT

The "As Is" analysis was performed to understand and document the current environment at Navistar. This was accomplished by studying the current organization and identifying the activities, systems, and information used by people within the organization. Preliminary analysis included preparing the organizational node chart, analyzing the business strategies, identifying existing systems, and defining activities and information requirements.

After this information was processed, we formulated the node chart and developed the matrices and the systems flow diagram. The purpose of this exercise

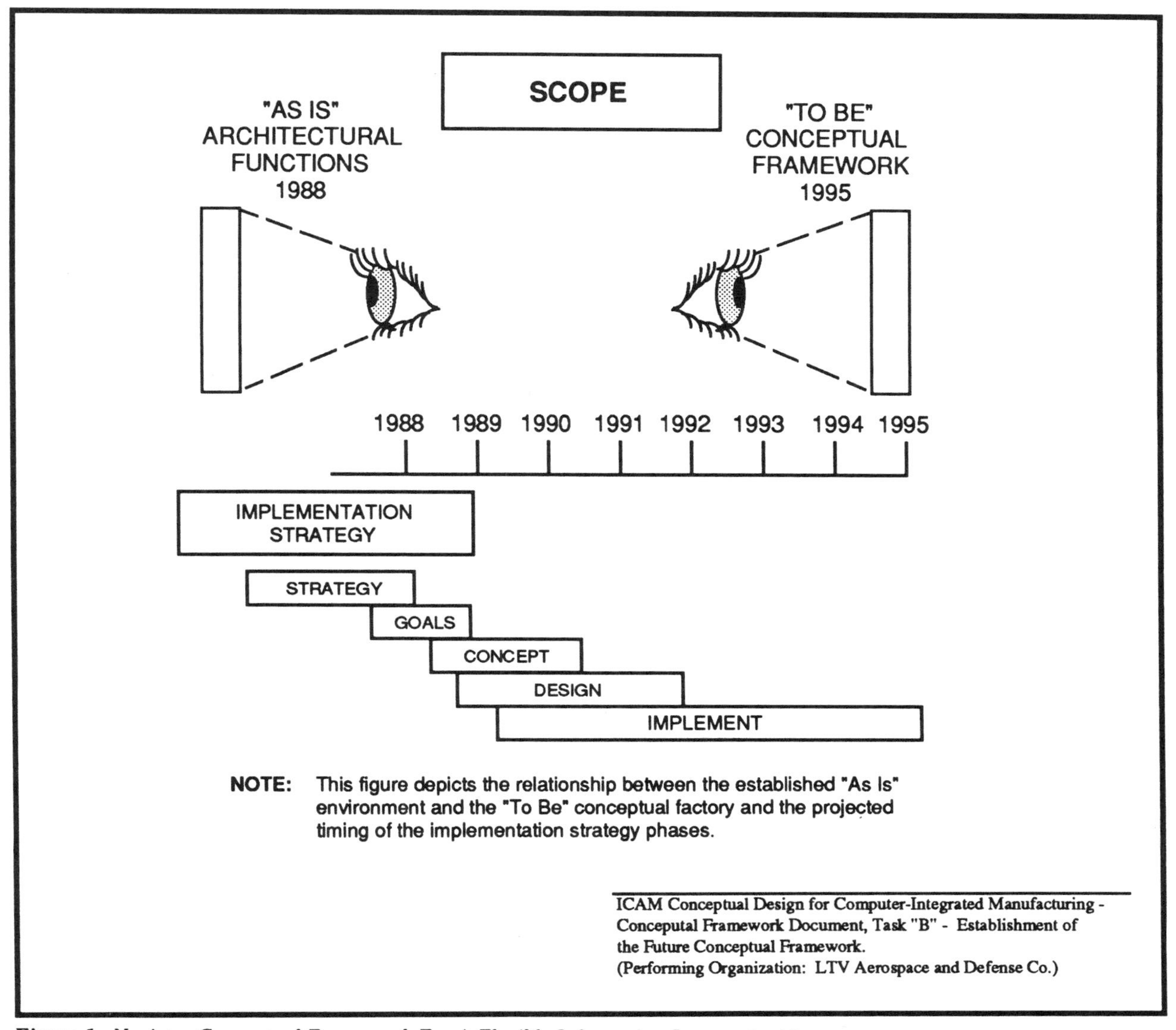

Figure 1. Navistar Conceptual Framework For A Flexible Information Systems Architecture.

was to summarize the characteristics of the information environment as much as possible so that we could understand exactly where we were and how things operated. This was not a statement of good or bad—just "here it is and how it works." At a minimum, the future environment must be able to accomplish the work as performed. It is important to see how the factory functions as a whole, not as a set of fragmented functions with specific control of certain information elements. Our objective was to look at the current operations and see how they work together. Only by doing this is it possible to look at the factory of the future and define its characteristics in light of what we know now.

The analysis was top-down to gain an understanding of what top management needed to know to get the job done and how subordinate functions (down to the shop floor) support top management objectives. The "As Is" defined the current coordination among the business units in terms of people, their roles, and their methods of operation with regard to information use. The "As Is" also pointed the way to the future environment in terms of needed changes or reorganization.

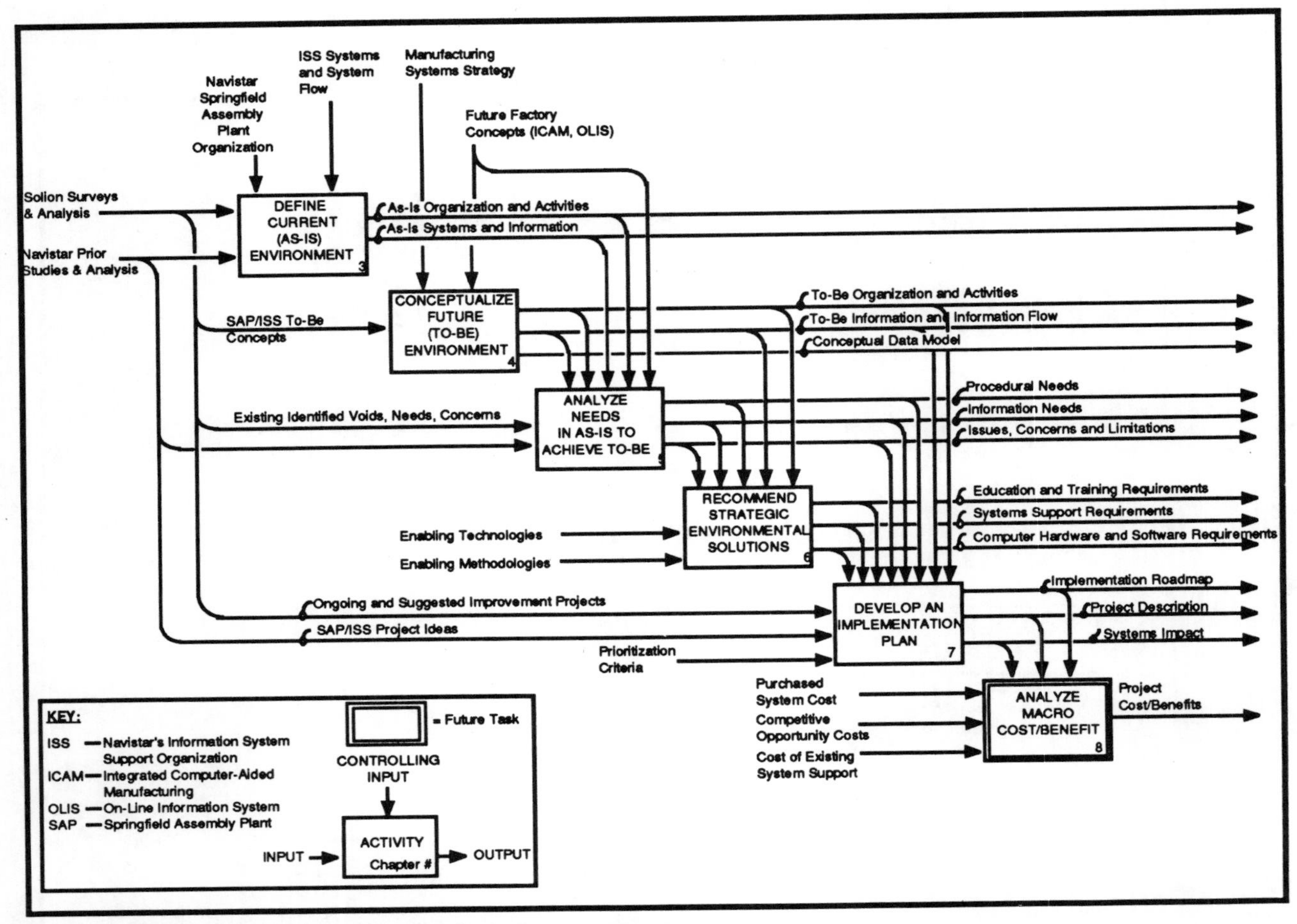

Figure 2. Distributed Systems Architecture Study Document Layout.

Current Springfield Assembly Plant Organization Structure

The Springfield Assembly Plant organization structure (shown in Figure 3) provided a general framework for understanding the activities of assembling a truck. It also provided insight to the general plant environment with respect to assembly, support functions, and information management.

The current organization structure has evolved to (and effectively does) assemble trucks. We cannot lose sight of this primary goal: plant support functions, engineering, ISS, and a major portion of corporate structure exist to support the profitable assembly of trucks. Our mission was not to critique Navistar's current success at doing this, but rather to improve upon it by providing more effective information communications.

Functional Activities Information Requirements

The following functions were identified based on the "As Is" Functional Node Chart (Figure 4):

- Planning,
- Industrial Engineering/Maintenance,
- Material Control,
- Production,
- Accounting/Financial Management, and
- Human Resources.

The "As Is" functions (or activities) relate very closely with the organization structure. This is because

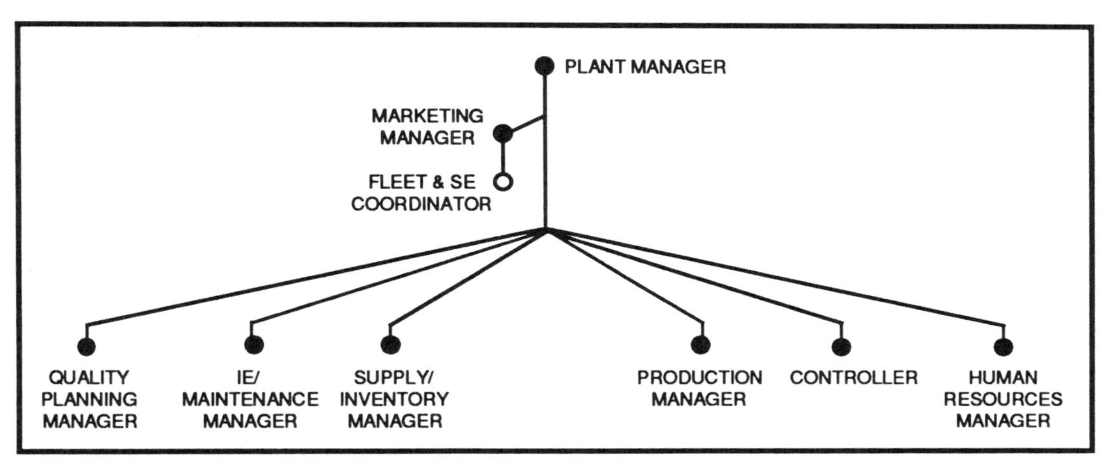

Figure 3. *"As Is" Organization of Springfield Assembly Plant (High Level).*

the organization structure is usually organized around the activities to be performed. Figure 4 was derived from the surveys with representatives from each major organizational unit at the Springfield Assembly Plant. At this level of detail, the actual duties/tasks of each area (e.g., planning, industrial engineering, materials, and production) are described and provide definition to what the staff in the organizational chart (as shown previously in Figure 3) are doing.

The general "As Is" flow (shown in Figure 5) began to define the information in and out of the major functions of the plant. Planning basically interprets engineering design and BOM applications to control the plant parts master record and production release of design. Planning also provides the process planning package, parts broadcast, and appropriations. Industrial engineering gets the process planning package from planning and performs a general manpower planning and routing. Material Control performs the scheduling and sequencing tasks in order to support material requirements planning and vendor ordering. In addition, Material Control is responsible for receiving, warehousing, expediting, traffic, and quality assurance for incoming materials. Production (assembly and subassembly functions) carry out the work defined by planning and industrial engineering to the schedule determined by Material Control with materials they provide. Station level supervisors interpret model-based instructions and truck specific special equipment requirements and part level production breaks, all within a time window not much larger than the line cycle time. The actual assembly steps are performed including the extra tasks of sign off and verify, problem reporting, and being trained.

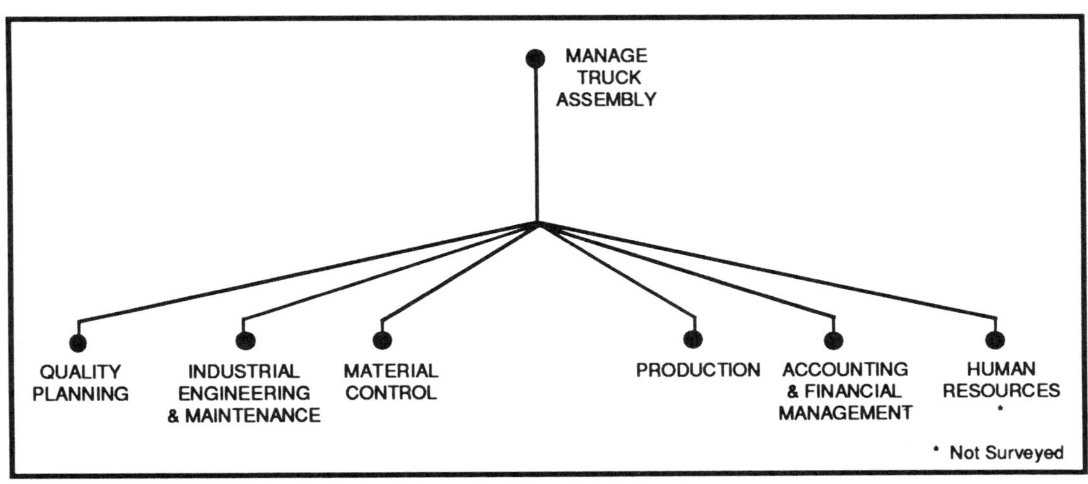

Figure 4. *"As Is" Functional View of Springfield Assembly Plant (High Level).*

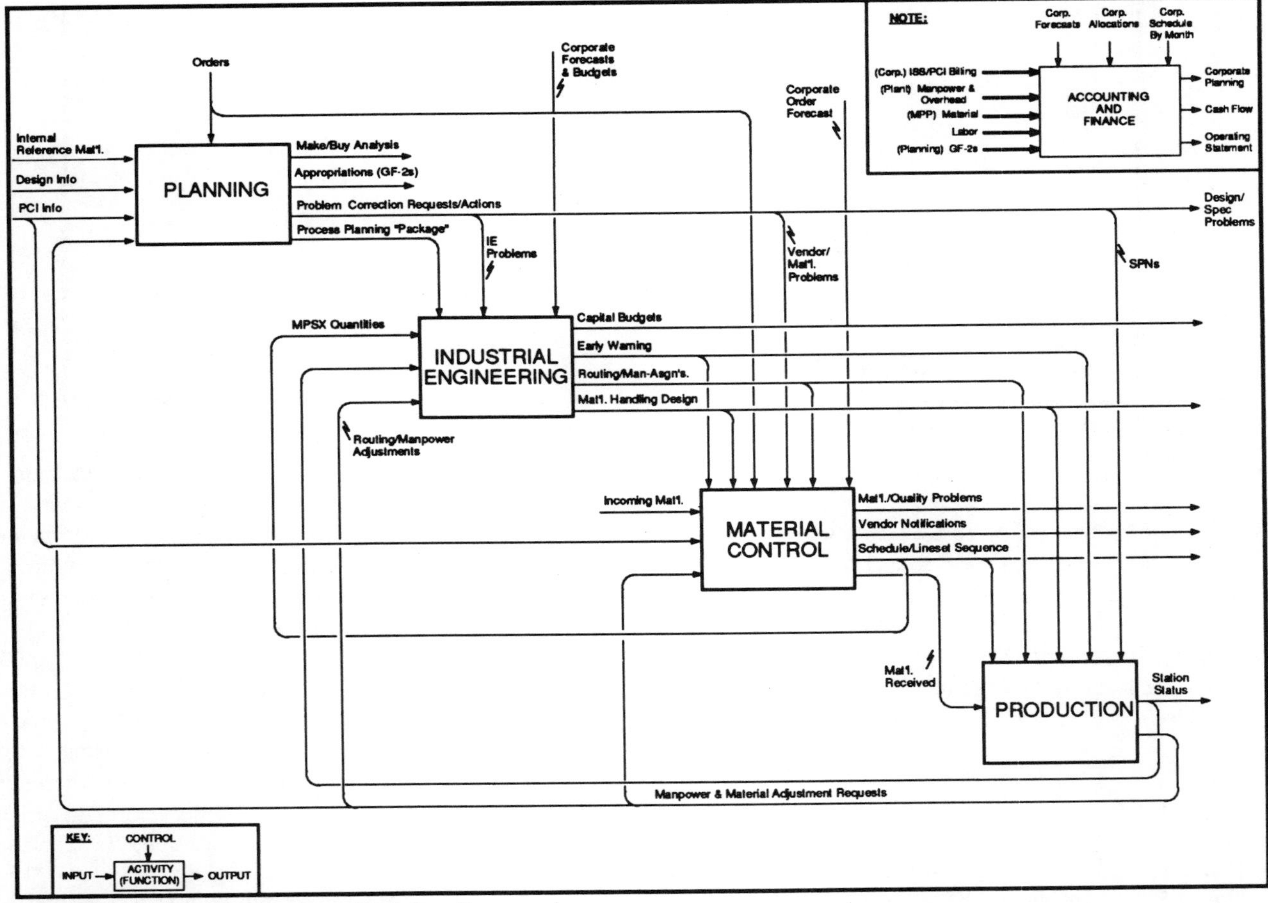

***Figure 5.** General "As Is" Flow (High Level).*

The bulk of the information burden rests on production's shoulders, not because they have the largest information requirement, but rather because they are where the information all comes together in order to build a specific truck. All other functions plan and support the production/assembly effort. The largest information coordination effort belongs to Material Control as they support assembly by planning and providing material delivery to the plant from a host of suppliers/vendors. The sheer explosion of information requirements at the materials level makes the accuracy of planning activities a critical factor. The importance of process planning, manpower requirements, and routing to effective assembly steps cannot be overstated. In other words, information accuracy and timeliness are a key resource to all the support functions if assembly is to proceed in an efficient and effective manner.

These plant functions are strongly influenced by a number of corporate level external control systems, such as Accounting, Purchasing, and Engineering, as well as systems which must interface to other Springfield Assembly Plant entities and other Navistar locations and suppliers (e.g., the Body Plant, the Fiberglass Plant, and the Engine Plant). This results in two additional functions:

- Corporate Interface.
- Supplier Interface (other Navistar plants and Vendors).

"As Is" Systems Environment

The Navistar Springfield Assembly Plant was characterized by a large number of information systems which had been designed to serve specific needs of individuals within the factory. The systems were large and complex and did not always supply timely and accurate information.

Corporate ISS is responsible for the design, programming, and maintenance of the major information systems for the Springfield Assembly Plant. These big applications typically run in batch mode each night at the corporate computer center. Local Springfield Assembly Plant ISS is responsible for printing the necessary reports at the plant, troubleshooting, problem reporting, PC and mini-computer applications, systems training, some limited TSO support, and computer communication support. TSO is IBM's time sharing operation but is used generically to refer to user written and maintained programs through RAMIS and other languages available for user programming. TSO programs are seldom supported by either ISS (plant or corporate) except when local troubleshooting may be required or if corporate ISS adopts a specific TSO program for support.

The present systems were developed over the years primarily as stand-alone applications, but each system is highly dependent on data generated from other systems. This means that the run sequence of application and the interdependency of applications is very important. This complicates major enhancements or replacement of critical applications because its interdependencies are not always well documented. In addition, several TSO systems, most of which are not formally supported by corporate ISS, are dependent on existing systems.

Another issue with the present stand-alone applications is user interface. Communication is either through manual interfaces or hardcoded applications written specifically to allow users to access information from one or more applications. This requires the users to either become proficient with each application or to find some resource to develop an interface between applications. The problem is further complicated by the fact that any change to the application causes changes in the interfaces and requires retraining of the users. Documentation quickly becomes obsolete and configuration management becomes increasingly difficult.

FUTURE ("TO BE") ENVIRONMENT

The objective of the "To Be" Environment was to establish an ideal future information systems environment, regardless of what it would take to achieve. The mission of the Needs Analysis (discussed later in this paper) was to explain the requirements to get from the "As Is" to the "To Be" environment. Also discussed in this paper are the recommendations on how to achieve the requirements and a project plan to implement the recommendations.

The "To Be" is shown in the same basic format as the "As Is" but is significantly different in that the environment is structured to meet customer needs and competitive pressures, regardless of today's organizational structure. The ideal concept for integrated information systems requires coordination to move functions closer together in terms of completeness and timeliness of information flow, provide more accurate data, better control, and faster execution.

The criteria that will distinguish Navistar from other organizations will be their ability to deliver a quality truck faster and better than anyone else. The "To Be" environment will allow Navistar personnel to:

- Process orders quickly and accurately;
- Engineer the specifications for the order;
- Coordinate the specs with manufacturing;
- Deliver accurate information to production;
- Process delivery and follow up quickly, and efficiently.

The only restriction placed on this concept of the "To Be" environment was to keep within the currently existing scope of the Springfield Assembly Plant. Corporate support functions, including purchasing, accounting, marketing, sales, forecasting, and engineering are assumed to be out of the direct control of the plant. Other "To Be" structures could be proposed which distribute some combination of these external support functions, but it would need to be coordinated with strategic manufacturing goals and objectives.

"To Be" Functional and Organizational Structure

The future "To Be" functional organization of the Springfield Assembly Plant (Figure 6) remained consistent with the "As Is" activities, except that it was rearranged into a "Plan," "Provide," and "Produce" structure. This was based on the structure formulated by a consortium of aerospace manufacturers for the Air Force Factory of the Future project. (Document title is included in Bibliography.) This closes the control loop within the plant to the highest degree possible without adding external corporate support functions (e.g., design). A control loop consists of activities which plan, provide, produce, feedback, measure, and adjust. The feedback, measure, and adjustment steps will be a function of plant management at a high level, and each department's responsibility within the guidelines of

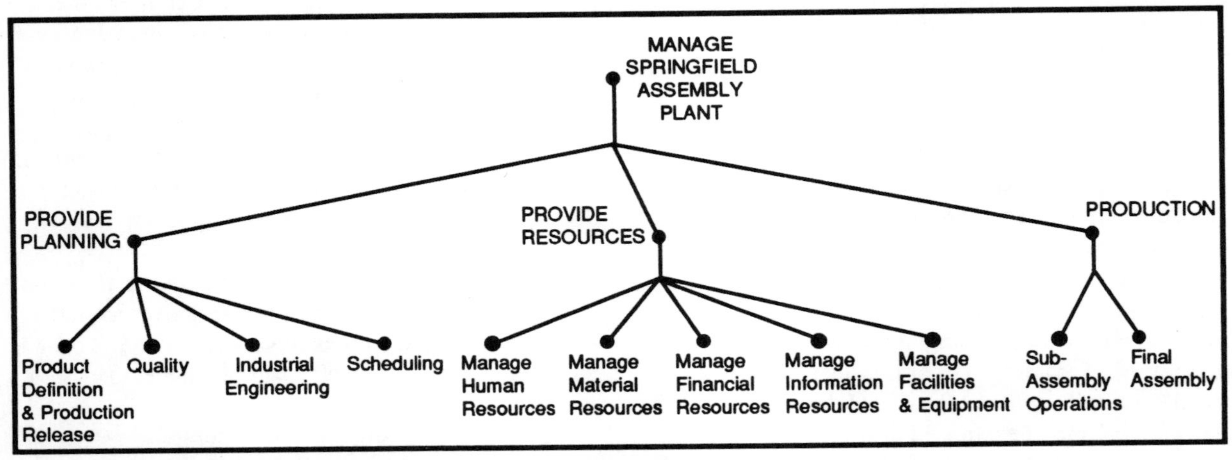

Figure 6. *"To Be" Functional View of Springfield Assembly Plant (High Level).*

plant management at a tactical level.

In the "To Be" environment, the functional structure of the plant might also become the organizational structure. The purpose and objectives of change need to be thoroughly understood by the entire organization. Policies and procedures need to be revamped to capitalize on improved information flow potential. One organization change will not satisfy all future requirements for the plant (e.g., the "To Be" is a dynamic structure), therefore, the process of change for improvement needs to be established. The primary method to bring new policies into use and set the stage for future change is training (refer to the Recommendations section for a detailed discussion). Finally, the information systems must support the dynamic organizational units and support the overall flow of communications within the plant.

Manage Truck Production

Managing truck production for the Springfield Assembly Plant consists of the same activities that are currently being performed, which include summary performance data analysis, exception reporting, and overall facility scheduling (e.g., production objectives, plant total manpower allowance). The "To Be" environment provides more decision support systems and ad-hoc query capabilities in order to improve decision making support. Also, the Plan, Provide, and Produce functions will have more thorough production status data to feed back to plant management for control purposes.

Provide Planning

The next few paragraphs highlight the planning, provisioning, and producing functions as conceptualized in Figure 6 ("To Be" Functional View) and in Figure 7 ("To Be" Functional Flow). This was our vision of the future environment, which would evolve in step with the manufacturing strategy.

The very heart of what will differentiate Navistar from other truck producers in the future is product and process information. Therefore, Product Definition and Production Release, Quality, Industrial Engineering, and Scheduling are all very important functions which will work in concert with each other. The primary information resulting from "Provide Planning" (see third box in Figure 7) consists of the manufacturing plan, the process plan/production (design/BOM) release, and the schedule.

Provide Resources

The concept implemented in the functional arrangement is that people, material, money, information, and facilities are all provisioning activities. They all need to be at the right place at the right time, according to plan.

One of the best ways to keep the entire organization remembering that they exist to produce trucks is to give them a piece of the action. Human Resources is a very good place to start. In addition to the common personnel tasks, actual manpower assignment, training, safety, and time and attendance gives a new meaning to human resource management as well as a critically important role in support of actual production tasks. If training/retraining the factory is a key to future improvement and modernization, then Human Resources will perform a vitally important role in developing skills, thereby assisting in the necessary cultural changes.

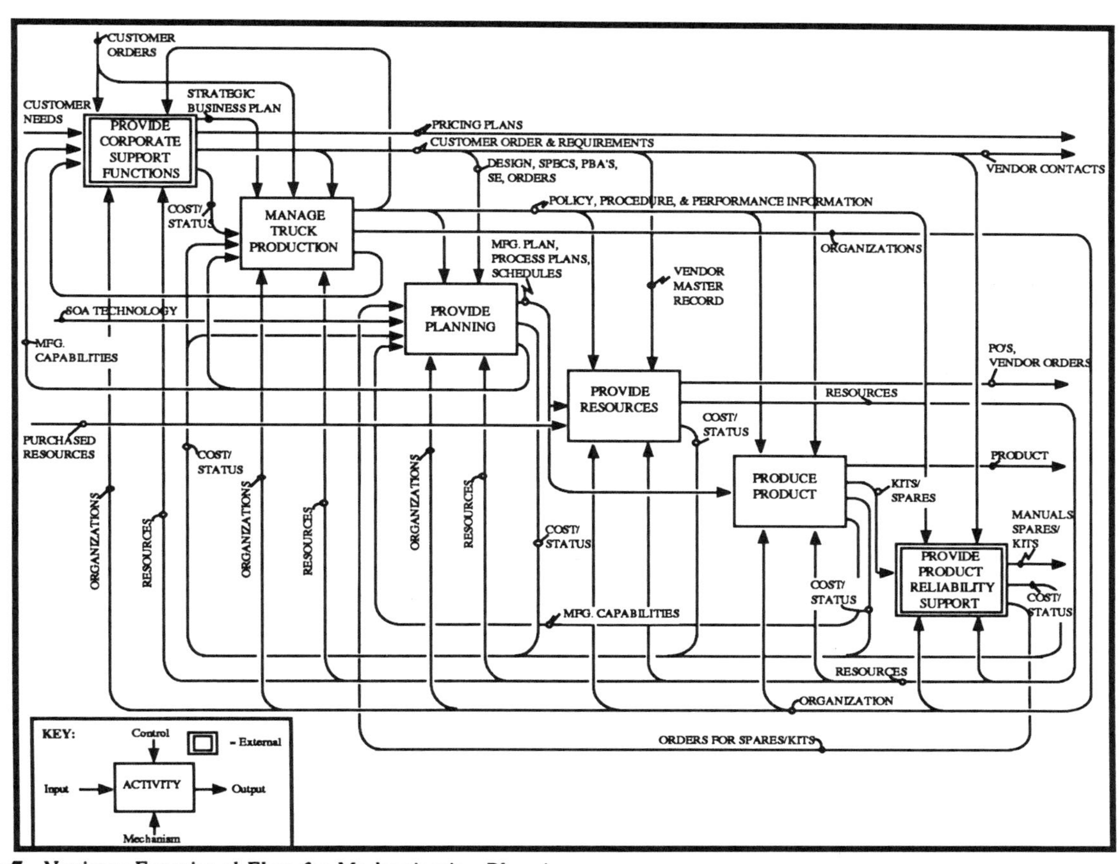

Figure 7. Navistar Functional Flow for Modernization Planning.

Information needs to be regarded as a plant/ corporate resource and be widely distributed throughout the plant (e.g., through ad-hoc reporting). The plant will need to have a coordinated information systems strategy which remains consistent with the corporate and manufacturing strategies. For example, if the manufacturing strategy aims for strategic business units (SBUs) which remain dependent (or as corporate and plant support functions), Figure 6 represents subassembly and final assembly steps as depicted. If the manufacturing strategy aims for independent SBUs (e.g., independent centers of excellence), then Figure 8 may be more accurate for subassembly functions. The autonomous SBUs require some redundancy of plan and provide functions, but also reduce the amount of detail needed at the plant level functions. Each SBU would then maintain its own bill of materials, process plans, inventory, and perform its own scheduling based on subassembly due dates and capacity. In any event, systems design and development should work closely with Quality, Industrial Engineering, and plant management to simplify procedures before they are automated.

"To Be" Information Flow

In the future, systems must be improved to remain competitive in truck manufacturing. They must be streamlined to a manageable size by taking advantage of existing and emerging technologies. If the systems are going to keep up with the growing demand for information, they must move to where the users are, placing capability and data where it is most accessible.

Manufacturing must learn to treat information as a company resource in much the same way as they handle and control material use. In this way, information is made available to everyone who needs it, when they need it. The future environment will rely on increased automation, the ability to transfer data electronically, and distributed processing to support the users.

The actual information (manual, verbal, and automated) used by an organization does not vary greatly by changing the organization structure. This means the information used in the "To Be" is very similar to the "As Is" information requirements.

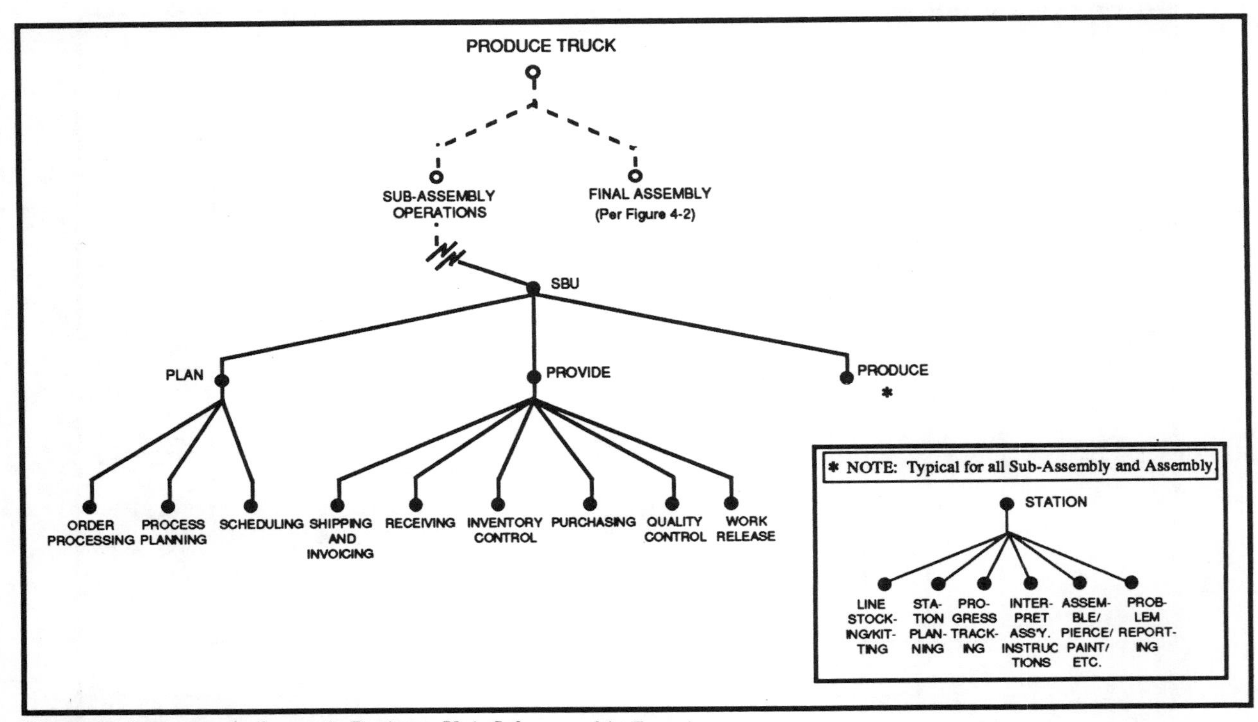

Figure 8. Autonomous Strategic Business Unit Sub-assembly Functions.

While the information used is fairly static, it has little or no bearing on the usefulness of current ("As Is") systems in the "To Be" environment. Information systems are very dependent on business rules/procedures and organization structure has a large effect on the applications. The "As Is" systems were developed one at a time with no specific manufacturing/information strategy targeted, and no overall information structure/model to work toward. This was simply the data processing approach used by most companies prior to 3-schema development methodologies (refer to the Recommendations section for more details). The future environment will need to accommodate change at a faster rate than the current one. Information systems will need maximum flexibility. The information flows in the "To Be" environment are dependent on a detailed manufacturing strategy. Similarly, a systems evolution plan (current systems revisions and replacements, or new systems) also depends on the manufacturing strategy. Therefore, the information systems strategy must be integrated with the manufacturing strategy.

NEEDS ANALYSIS

The needs analysis section was included in the study to bring together the "As Is" and "To Be." It defined the problems in the current environment and the actions that must be taken if Navistar is to proceed with the plan to achieve the "To Be" environment. In the analysis, it was necessary to identify the voids and deficiencies in the current environment so that the improvement projects could be understood in the context of the future.

Figure 9 graphically represents the process of needs analysis with regard to the Springfield Assembly Plant. Our objective was to concentrate on information specific needs and identify specific information

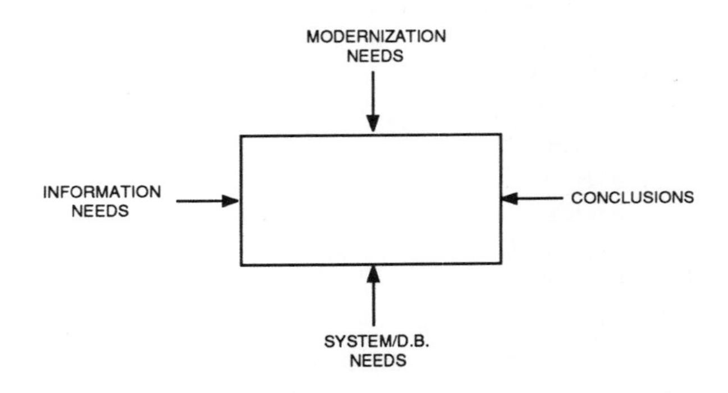

Figure 9. Needs Analysis Process.

systems projects which needed to be performed to support an "Integrated Utilities" strategy within Navistar. The existing systems and databases serve as the means of identifying what data is where right now and, therefore, how to make the data available to those who need it.

Information Needs—User Perceptions

As stated above, the information needs were derived from a number of sources. One of the most informative sources was within the context of interviews with specific organizations in the plant. This included Planning, Industrial Engineering/Maintenance, Production, Material Control, and Accounting/ Financial Management. Each group felt that group input to design and better communication were critical to improved productivity.

The custom nature of most trucks and the ability to introduce change late in the assembly process are important issues. It highlights the need for a Flexible Information Systems Architecture and possible suggestions in some areas where policies and procedures need to be reviewed. A "flexible" system is still dependent on having good, clear policies and procedures in place. Automation seldom provides a solution to an undefined process.

Potential "To Be" Objectives

Since the "To Be" is a targeted ideal environment, it is useful to list possible objectives which answer the question "what would we like to accomplish?" Table 1 is a list, from general to specific, of ideas for the future information systems environment. This is the starting point from which to formulate needs, recommendations, and project plans.

Flexibility

Both the existing systems environment and the target environment evolve in time (Figure 10). What is being presented in this document is the result of the first iteration in defining the future concept. Part of what creates the changing concept of the target environment is the changing state-of-the-art technology and changing manufacturing strategy. Likewise, implementation of any new technology causes change in the existing system.

Table 1.
Potential Information Systems Objectives

OBJECTIVE	OBJECTIVE
Simplify Procedures and Systems	On-Line Labor Routing
Integrated Multi-Vendor Environment	Improved Labor Standards
Data Consistency	Effective PBA Decision Support
Integrated Data Collection	Peg Requirements To Model To Specific Trucks
Training (Ongoing)	Peg Inventory To Specific Operations
Increase Computer Processing Window	Manpower To Specific Operations
Decrease Computer Processing Time	Track Progress At Station Level
Management Information System	Manpower Simulation
Auto Data Entry	TODB Change Notification
Integrated Scheduling	Better S.E. Handling
One Forecasting System (Mat'l., Labor)	CBT (Computer Based Training)
Integrated EPL and Drawing (out of scope)	Preventive and Predictive Maintenance Tracking
C.A.P.P. (Labor and Material Routing)	On-Line Library of Reference Material
Dynamic Materials Requirements Planning	Problem Reporting (On-Line?)
Single/Enhanced B.O.M.	Vendor Monitoring/Evaluation
Real Time Assembly Aids	Serial Number Component ID To VIN (Vendor Identification Number)
- Parts to Station, Parts to Person, PBA versus Routing	Part Use History (As Built Truck B.O.M.)
- Cribs, Supervisors (Drawings)	Line SPC (Statistical Process Control)
- Related to Truck on the Line	
- Look Ahead - 3 Trucks to 5 Days	
- Part Substitution	
Quality/Performance Measurement	
Batch Parts Master	

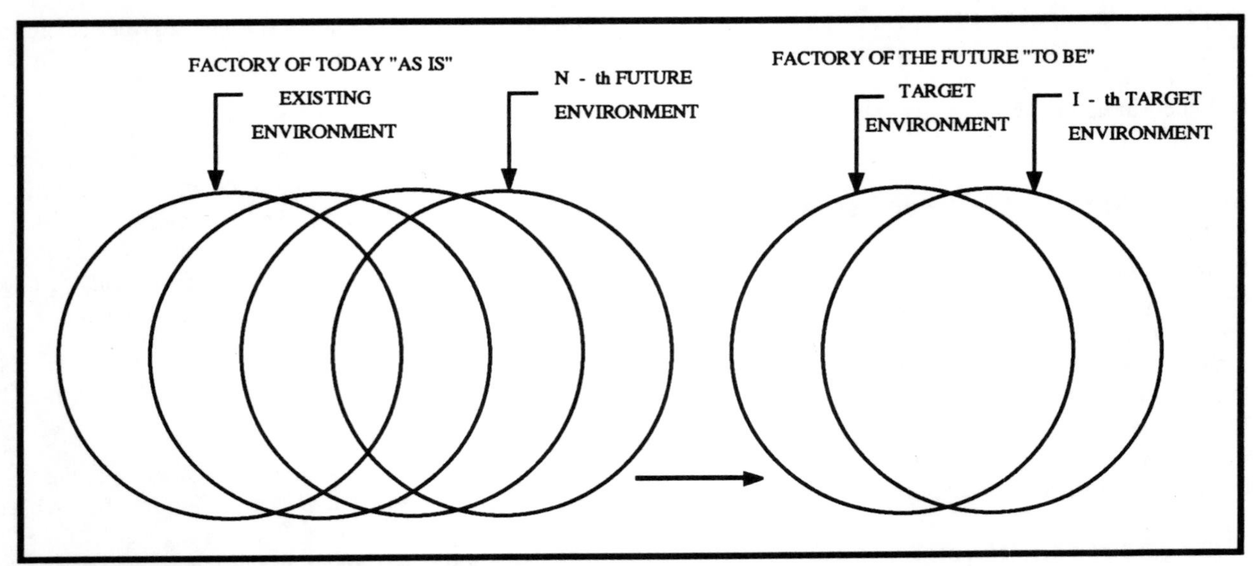

Figure 10. *Systems Environment Evolution Process.*

Manufacturing strategies will change based on market conditions and changing management philosophies. The strategic changes effect the most change in the organization, policies and procedures, and, therefore, create the largest impact to systems supporting the enterprise.

Information systems will, therefore, need to be expandable, flexible, and integrated using state-of-the-art technology to its fullest advantage. The technologies include hardware and software which streamline the process of doing business. Using integrated databases at the heart of the operations can assure accuracy and accessibility. The trend for the future is to establish an enterprise view of the data so that all users have ready access. It is a competitive requirement to begin a path to integration which will encourage and support the integration of information across functions in the company.

The managerial techniques required to support this flexibility include opening up lines of communication, involving everyone in the process, training extensively, and reorganizing/modifying/adding functions where necessary. Modernization becomes everyone's responsibility from the president to the operators on the factory floor. People are expected to take responsibility for what they do. Failure is not expected, but it is accepted. Producing increasingly complex trucks faster raises the risk of failure on occasion, but if improvements are going to be made then experimentation becomes a way of life. Training everyone all of the time becomes important if employees are to understand what is happening and why it is happening. It also makes employees feel that their company values them and their contributions to the company.

System Evolution

Over the course of the next three to five years, many of Navistar's current systems are going to be modified, enhanced, and/or replaced and entirely new systems will be added as well. This will happen as a regular part of system attrition, regardless of approach.

Purchased software packages will be a good solution when they meet the requirements of the intended user base and can be integrated/interfaced to existing systems environment. There are two main reasons why the overall systems plan is important in the decision: (1) ease of use, and (2) information accessibility.

Ease of use translates into ease of input—can the purchased system work off of data already being collected and processed in other systems? This helps prevent a redundant data entry problem. Information accessibility is important for sharing information with other systems which require it and preventing unwanted data redundancy. This information structure needs to be defined, available, and queriable in the future to achieve overall system flexibility and availability to all relevant users.

A strategic plan plays a critical role in evolving the overall information base into a more flexible whole by establishing an organized and consistent set of goals, objectives, plans, procedures, and standards. Otherwise, development will continue to occur on an application-by-application basis and the systems will get more and more intertwined, complicated, and eventually fail.

RECOMMENDATIONS

The recommendations in this section are based on satisfying Navistar's immediate corporate objectives and long term business strategies. The implementation of the recommendations presented here should show direct benefits in terms of Navistar's ability to introduce new products quickly. The recommendations are aimed at three main issues: cost, quality, and schedule. Key features of Navistar's factory of the future will be integration of business, engineering, and manufacturing data and the creation of an atmosphere of increasing innovation for manufacturing and information systems. The combination of the two should help to create an excitement and willingness to make something happen by creating an architecture to stimulate change.

Systems Environment

Navistar has currently in place a large investment in a variety of hardware and software which the company can ill afford to replace all at once. These systems work together in a semi-automated fashion with some data redundancy. A diagram of existing systems portrays a complex computer systems environment that is vital to Navistar's daily business needs.

The manufacturing systems of the future must support access to distributed information across a variety of subject databases. These databases will support a variety of users throughout the company. Distributed data is important since it must be quickly accessible by all users. The compute power of current and emerging hardware makes it more efficient and cost effective to use smaller networked systems rather than very large central systems. By distributing data and functionality, the systems are not nearly as vulnerable to hardware failure. Even if one machine is unavailable, other resources can be used. Another advantage of distribution is the ability to focus functionality in the area where it is needed. Computers placed in the hands of users make specialized applications (such as CAD, financials, planning, and scheduling) operate more effectively since users do not compete for computer time and specialized equipment. Distribution gives the computer user the power of a personal workstation in a network with access to host computers. Engineers, planners, and schedulers can develop information locally and put it on the host when it is ready for general access. Controlling the access of "working" versions on a central host requires much more version control and access restrictions. Distribution can lead to simplification and streamline of process and procedure by allowing overlaps or certain redundancies.

These databases support a variety of users, both internal and external. The purpose of distributing the databases, as discussed above, is two-fold: performance and security. Performance is an issue since it is not practical to support all geographic users on-line from a central system. Response time would degrade and result in low user satisfaction. It is more practical to place an application database with the primary users. Access privileges are also an issue in that only specific users should have the ability to view, change, or create certain pieces of information (e.g., financials, drawings, material receipts). Individuals should be able to rely on the integrity of the data they own.

Typically, MIS and CIM specialists have had different viewpoints on computing because the first is concerned with business applications and the latter with manufacturing processing. Even when both speak of distributed computing, there is confusion about the applications which are being implied. In Table 2, the two viewpoints are merged with respect to the Springfield Assembly Plant.

A vision of the future is essential to hardware and network planning. A depiction of this vision is shown in Figure 11. Here, applications which have been identified in the needs analysis are shown on different computers based on our experience and current industry practice. The final choice of what level of computer an application runs on, however, is dependent on several factors, such as required response time, required up time, number of users, licensing fee arrangements, required support, amount of data, commonality of data, uniqueness of the application, and so on. All of these factors should be considered prior to selecting the appropriate computer on which the application will run. Furthermore, it is essential that the future environment be kept in mind for hardware and network planning.

Education Issues

Education issues are many and varied. Opportunity for change is probably greater here than in almost any

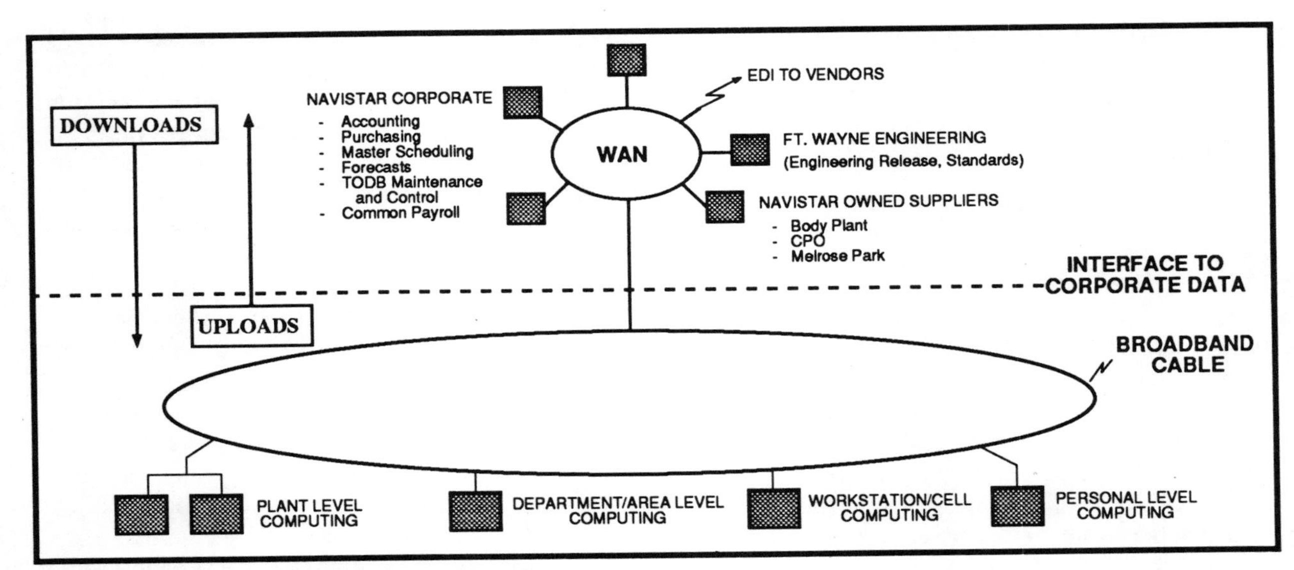

Figure 11. *Conceptual Hardware System Architecture.*

specific set of computer applications. Most immediate is the on the job training for line personnel. Many quality and process problems could be averted by providing assembly aids to line personnel. The net effect will be felt in the bottom line in terms of reduced cost, improved quality, and on time delivery.

Introduction of information technology at the Springfield Assembly Plant will require training at all levels from end users to developers. Training must begin with senior management so that they understand and support the risks and benefits associated with the introduction of new technology. Once management training is complete, the technical staff can be trained.

The technical staff (developers, support personnel, operators), depending on their responsibilities and specific assignments, will be trained on a variety of topics, including overall strategy and specific technologies (hardware and software).

End users must learn a variety of things depending on what experience they have and what applications they will use. The end users may require instruction in computer basics as well as specific applications.

To minimize "culture shock" it is best to prepare a continuous training plan and bring people along at their own pace. Self-instruction manuals and com-

Table 2.
MIS and CIM Viewpoints with Respect to Springfield Assembly Plant

LEVEL	CIM (Mfg. Process)	MIS (Business Appl.)	POTENTIAL PLATFORMS
Plant	DEC Based, Primarily ENG related	IBM Based Business Applications	VAX 6500, 8800, IBM 3090
Department/Area	Automated Mfg. Host	Department Computers	MVAX, IBM 9370, AS 400
Cell/Workstation	Automated Cell Host Engineering WS's	Decision Support	MVAX, SUN
PC/Controller	Robot controllers, AGV Controllers, PC	Miscellaneous PC Tools	IBMs, MACs, GMF
PLCs/Terminals	PLCs, Intelligent Sensors	Dumb Terminals	VT 320s, 3174s, AB PLCs

puter-based training are alternative delivery mechanisms that should be considered. Training should be conducted for everyone all of the time.

Resources

It is critical to the success of any modernization effort that people, money, and equipment be made available as firm commitments. Any major project requires participation by key plant personnel, whose time is at a premium. As a consequence, many systems' projects go without a plant sponsor and are, therefore, destined to failure before they begin. Top management commitment to modernization efforts is critical to creating a vision for the future. Since a culture change is needed, leadership by example, including commitment to resources and project plans, is key to motivating plant personnel. They need to believe in it.

Presentation of the overall modernization plan to all levels of the organization is needed. This will allow top management to gage the funding requirements and to commit to funding at a certain level, based on a proper business decision for the Corporation. This, in turn, will determine the speed at which the plan can be executed, which can then be communicated to plant personnel. Knowing where they are heading (and how fast) will eliminate much frustration. In addition, the plant must identify and commit the appropriate resources to each individual project in the plan. If this cannot be done, alternatives (additional staff, consultants, or another project) should be pursued and the impact to the plan communicated.

Approach

Due to the complexity of the existing systems environment, interaction with other corporate functions, and the magnitude of change from the "As Is" to the "To Be," an evolutionary approach must be followed. This allows production to continue while modernization efforts are under way.

It is also recommended that this modernization plan be input into an automated tool (CASE), updated on an ongoing basis, and used as a communication vehicle between different disciplines and corporate functions. It is necessary to see the whole picture, how each piece fits, and who is doing what (and how fast) in order to gain consensus and move to the fastest practical pace. Expected progress can be measured against the plan to ensure that things are happening at the appropriate pace. Issues which are currently not surfacing will come up as a result of having the ongoing plan. Also, as changes occur in the "As Is" and "To Be" they can be reflected and communicated appropriately.

The recommended approach then is incremental: small improvements, large improvements, and funding all within the context of an overall, published plan.

IMPLEMENTATION PLAN

An implementation plan was defined to evolve the factory to the "To Be" environment by prioritizing projects which were consistent with recommendations and responsive to the needs. All of the prior analysis had been done to formulate an effective implementation plan.

The goals for modernization, the technology which will be contained in it, and the key needs which will establish the basis for the conceptual framework have been discussed previously. Perhaps even more important than the actual technology that will be implemented is the mechanism by which change will take place in the factory. Only in rare instances will the opportunity exist to build complete new factories from the ground up. In most cases, the modernized factory must be introduced in an evolutionary fashion. As Navistar evolves toward the modernized factory at the Springfield Assembly Plant, technology will undoubtedly advance, making the evolution a never-ending process. Therefore, there must be an incremental implementation strategy for modernization.

For an implementation plan to be effective, it needs to be addressed from three levels: corporate, plant, and information management. The interests/concerns are common and unique to each. Each depends upon the others for success. The corporate manufacturing strategy is the driving force since the manufacturing strategy guides the implementation of information systems.

Improvement projects can exist which are necessary for any alternative future strategy. However, there is a decision point which is critical to the modernization of information systems, beyond which changes depend greatly on which strategy is taken. In Figure 12, the strategies are listed in order of difficulty (e.g., "A" requires far fewer enhancements/replacements to existing systems and fewer new systems than "E").

The plant strategy depends upon the corporate strategy. The organization must evolve to support corporate goals and the systems must evolve to support the organization. Technology to manage change and,

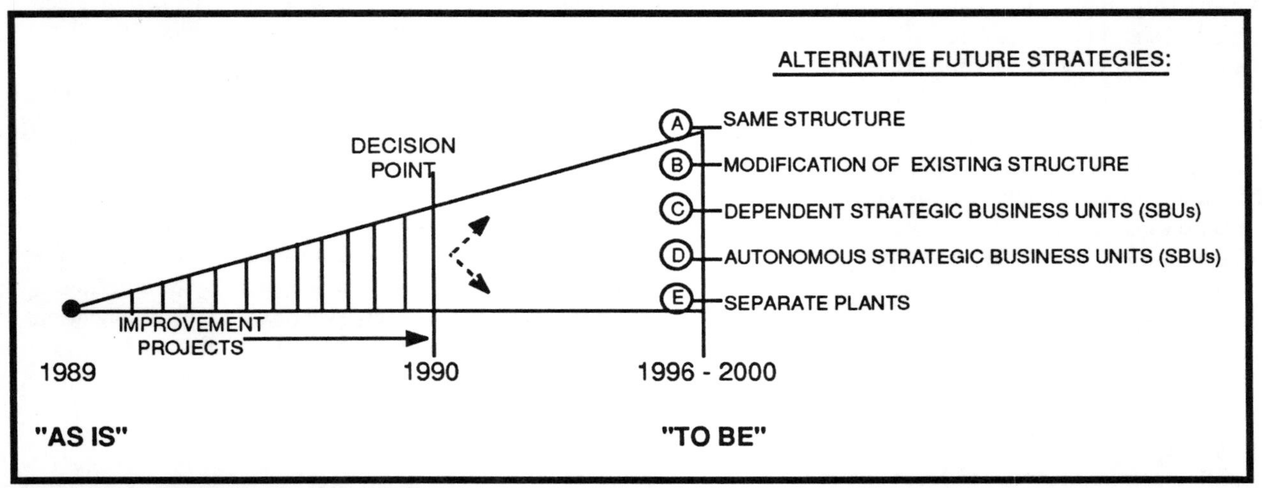

Figure 12. Strategic Manufacturing Objectives.

at the same time, support production must be put into place and skilled personnel must be trained at the plant to support the technology. The plant strategy cannot exist without the corporate strategy.

Implementation of Strategic Issues

The ability to adapt to change is part of the FISA strategy putting technologies in place to assess impact and maintain control. The information management strategy responds to corporate and plant needs with tools and information in a support role.

The plan defined for Navistar established a first cut rating of modernization projects based on:

- Importance - Importance related to achieving the "To Be" objectives.

- Phase - Timing issues with respect to sensitivity and importance.

- Sensitivity to Corporate Strategy - Highly (H) dependent on the future manufacturing strategy, Medium (M), or Low (L). This includes what effect each project has on related systems, procedural issues, availability of purchased systems, and duration of the project.

The projects reviewed were all necessary, but some were needed immediately. It is acknowledged that prioritization will change over time and the plan will be updated.

The immediate solution was to categorize all projects into a five-phase approach and an estimated duration. The five phases represent when a project could begin and the duration, and how long it is expected to last from start to finish. Not all Phase I projects were expected to begin in 1989. Some may not be started until 1994, for example—all depending on resource availability and manufacturing strategic objectives. The general meaning behind each phase is:

- PHASE I PROJECTS:
Immediate. May be critical to begin as soon as possible, or simply can begin. The projects are usually characterized by a low sensitivity to the "To Be" manufacturing strategy.

- PHASE II PROJECTS:
These projects tend to be more sensitive to future manufacturing strategies and/or are dependent on Phase I accomplishments.

- PHASES III, IV and V PROJECTS:
These are mostly dependency issues on earlier projects, technology, and/or highly manufacturing strategy oriented.

The manufacturing and information strategies are not expected to be in complete synchronization until Phase II, but many improvements can begin prior to that point and, in some cases, in anticipation of it.

The projects were ranked by overall importance to the "To Be" functional architecture and separated by phase. The purpose was to establish a guideline for the assignment of resources once the project scope had been detailed and the level of effort had been committed to by management (corporate and plant).

Importance was categorized by "A" - very important projects; "B" - important projects; "C" - useful enhancement/support projects; and "D" - enhancement/support projects which would be useful if time and resources permit.

SUMMARY

The use of analysis methodologies to organize and communicate facilitated the understanding and acceptance of the study. Participation by key individuals from the plant was invaluable to the study's successful completion.

BIBLIOGRAPHY

Integrated Computer-Aided Manufacturing (ICAM) - Architecture Part III -Volume IX—Composite Information Model of "Manufacture Product" (MFG1) (Performing Organization: SofTech, Inc.) June 1981.

ICAM Conceptual Design for Computer-Integrated Manufacturing - Conceptual Framework—Document, Task "B" - Establishment of the Future Conceptual Framework (Performing Organization: LTV Aerospace and Defense Co.) February 1984.

Navistar Information Flow - U.S. Truck Assembly, July 1985.

Navistar Application Transfer Study, June 1987.

Navistar Manufacturing Systems Strategy Overview, October 1987.

Navistar Manufacturing Systems Strategy Update, April 1988.

Navistar The ISS 3-Year Strategy, July 1988.

On-Line Information System (OLIS) White Paper (Authors: Dave Judson and Carol English) November 1987.

Needs Analysis Document for Navistar Modernization (Author: SAIC, Solion Division) July 1988.

Section Two:
EIX and Integration Basics

Section Two papers are:

- *EIX Definitions and Framework.*
- *Roadmap for Enterprise Integration.*

Various organizations' EIX definitions vary significantly since no industry standard definition exists. Therefore, to ensure a common understanding of terms and concepts, Groesbeck provides the latest accepted EIX and related terms' basic definitions. In addition, he illustrates the relationship of PDES, IGES, EDI, and CALS as EIX subfunctions. Groesbeck provides a four-stage framework to calibrate the stage at which an organization is positioned. The definitions and framework allow individuals to improve their communications, analyses, and planning efforts.

Mayer and Painter provide a framework for systems integration to aid the systems planning, development, and implementation efforts. Selection and use of methods and tools should relate to the nature of the organization, data, and technologies involved. This cornerstone paper provides a basic roadmap for Enterprise Integration, which is especially valuable in the systems planning stage.

EIX Definitions and Framework

HOWARD E. GROESBECK
Groesbeck and Associates

Today's business enterprise is typically a complex organization with extensive information being exchanged both externally and internally. An enterprise context diagram, Figure 1, illustrates some of the external bidirectional information exchanges involving customers, suppliers, stockholders, regulators, and other stakeholders.

Internal information exchanges involve traditional enterprise activities, such as marketing, sales, engineering, manufacturing, finance, logistics support, and human resources. As competitive pressures and operational complexity increase, the importance of quality and timely exchange of relevant information increases. Understanding EIX (Enterprise Information eXchange) characteristics, potentials, and implementation processes become considerations to astute businesses. To begin a basic understanding of EIX, a few definitions are useful.

DEFINITIONS

Information and data are related to each other but are not synonymous terms. For this document, data is defined as the undigested or raw facts that are available for organization. Examples of data include, but are not limited to, tolerances on a machined part, quantity of items in a box, actual component performance characteristic achieved, number of employees in a category, etc.

Information is defined as organized data made useful to someone or some group. Examples of information are tolerance trend line on a Statistical Process Control chart, certification that the delivered

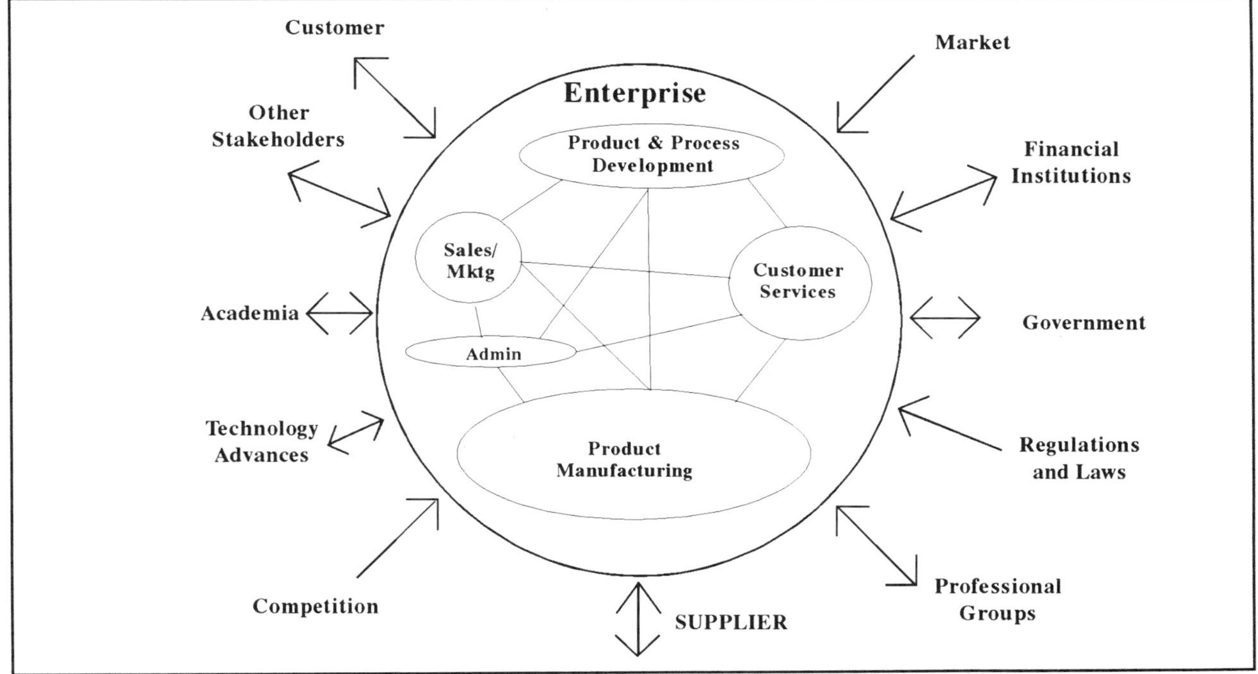

Figure 1. EIX Context Diagram.

goods are those ordered, performance characteristics described as meeting specifications, a table comparing the number of employees in various categories, etc.

This document focuses primarily on information. Whenever the term information is used, it assumes that the associated data elements are a part of the total information content.

EIX includes both business and technical information, exchanged both internally and externally.

Business information refers to all the information and data that is needed to operate the business. Typical business information includes customer, order, cost, price, employee, etc.

Technical information is the detailed information and data used to develop, manufacture, and support the product or service. Typical technical information and data includes research, engineering, manufacturing engineering, product or process specifications, plans, configurations, designs, analyses, documentation, quality, testing, reliability, maintainability, and/or logistics information involving alpha-numeric, geometric, image, audio, or video content in paper, intellectual, and/or electronic formats. Initially emphasized technical documents included product drawings, technical manuals, process plans, service/maintenance/repair technical manuals, and logistics support technical data.

Technical Data Interchange (TDI) refers to the exchange of a business's technical information within the organization and externally to the organization. TDI's initial emphasis was on externally exchanged information, but does apply to internally exchanged technical information.

Information and data can be exchanged intellectually (orally or sign language), physically through paper, and/or electronically through computerized input/output, processing, and storage devices.

Information exchange typically implies that a copy of the source information as a package is handed from one source to another. If the information is not in the correct format to be understood by the receiver, it is translated into the correct format. In this technique, two copies of information exist often creating control and synchronizing problems.

Information interface refers to a configuration in which one computer system directly ties into or interfaces with another computer system. This is accomplished periodically in a preagreed process and format when a transfer of information is needed.

Integrated information sharing is achieved when two separate systems are designed to be interdependent of each other's processing and information sharing. Information is maintained in only one logical location and is interactively accessible whenever needed, giving the appearance of one singular cohesive system.

Recently, the manufacturing enterprise exclusively exchanged both business and technical information and data in paper format or orally. This process was time consuming, costly, and prone to errors. The use of computers provided additional efficiency to the information dilemma. However, due to the multiplicity of different computer applications and systems, this improvement was only a partial solution. Leaders then looked to standards-oriented computer-based electronic transfer as a way to enable dissimilar computer systems and applications to be able to exchange information. The following identifies the major standards.

STANDARDS

The principal standards established to improve productivity, lead time, cost and quality of information exchange are IGES, PDES, EDI, and CALS. Each of these programs will be briefly introduced here but moreso in other sections of this book. For a comprehensive understanding, these programs should be thoroughly investigated by reading the materials in this book and those listed in the reference bibliography.

IGES

The Initial Graphics Exchange Specification (IGES) was established in 1979 by a group of large aerospace manufacturers for internal technical mechanical product geometry data exchange between differing computer-aided design and drafting systems. This de facto standard was helpful in improving internal and external information exchange efficiency. However, IGES had many problems limiting its use. IGES transfers initially were accomplished by magnetic media exchange between organizations. More recently, direct exchanges via computer networks are used.

PDES/STEP

The PDES/STEP (Product Data Exchange Specification/STandard for the Exchange of Product data) evolved from the initial PDES standard, which was established in 1985 to replace the problematic IGES de facto standard. The PDES/STEP standard is structured for the exchange of technical information needed to design, manufacture, and support a product including geometry, tolerances, geometry, related models, numerical control, materials, and structure information.

EDI

Electronic Data Interchange (EDI) has been successful over the last 10 years in the transportation, chemical, earth-moving equipment, retail, banking, and automotive industries involving primarily external business information and some technical information exchange. Business information, such as invoices, purchase orders, purchase order acknowledgements, advanced ship notices, quality data, and bills of lading have been successfully electronically exchanged internationally by the EDI capability. EDI is primarily accomplished through controlled access data transfer using a secure and neutral third party computer system as a form of "post office." EDI also includes directly linked computer systems, but due to many technical and security reasons, is not used very much. Many early EDI programs resulted in a shorter transaction time, but at a higher net operational cost. The total cost impact should be contrasted to the benefits of time savings to avoid implementing "technology for technology's sake." Today many large organizations, a few of which are identified in this book, are successful using EDI to shorten cycle time and reduce operational costs.

CALS

The Department of Defense (DoD) established the Computer Aided Acquisition and Logistics Support (CALS) program in 1985 to standardize the technical information exchange between weapon system contractors and the DoD. The DoD, in dealing with a large number of suppliers, was in effect causing dissimilar systems information to be converted into paper and then reconverted into electronic format for government use. The DoD paper process and cost was unacceptably excessive. The CALS program is the DoD's strategic program to improve weapon system lead times, costs, and quality.

STANDARDS MODEL

To further illustrate the general applicability and relationships between the four standards and a manufacturing enterprise's value chain, Figure 2 is provided as a model.

CALS provides the most extensive technical information exchange coverage. However, as a

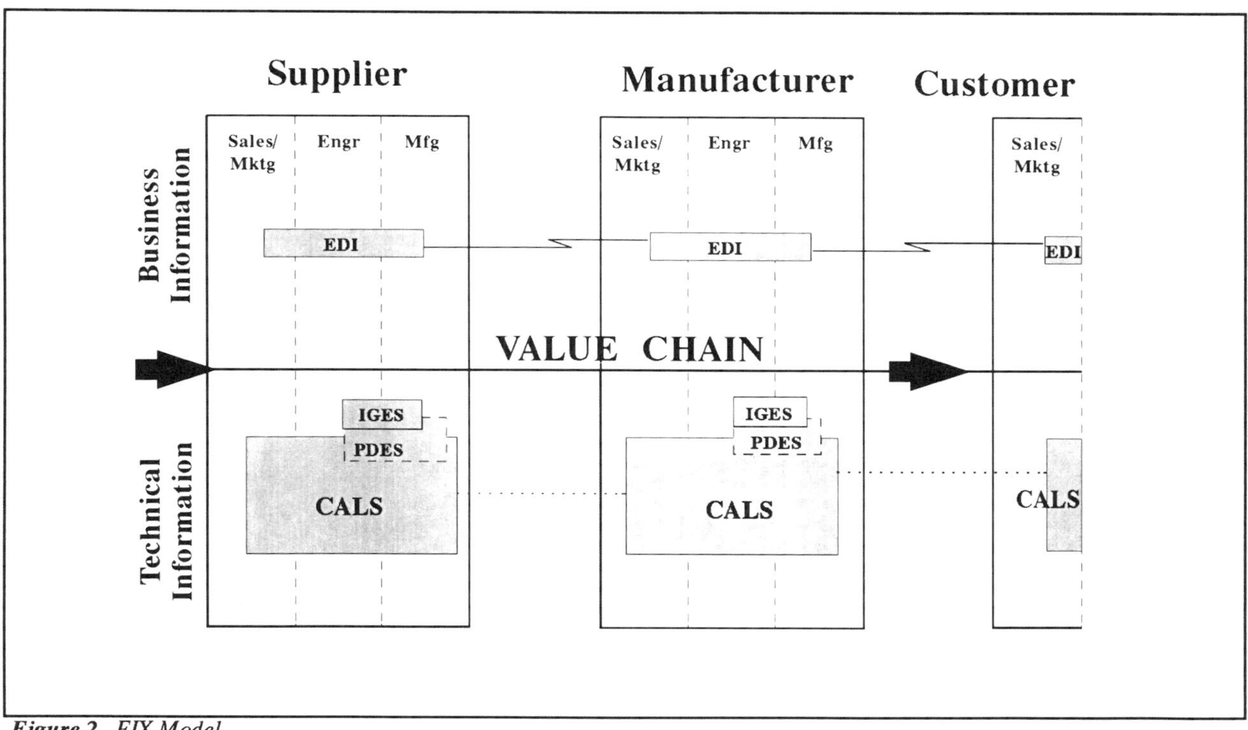

Figure 2. EIX Model .

standard, it is not complete at this time. EDI provides coverage of discrete business information. Lastly, PDES and the older IGES provides some valuable exchange of product geometry and related development information within and externally to the organization. What clarifies the overall direction of EIX is a basic framework described in the next section.

FRAMEWORKS

A four-stage framework depicting the evolutionary state in which EIX is progressing is provided in Table 1.

The paper and/or intellectual information exchange, according to Muglia (1992), was characteristic of Stage 1 information advancement level. Ultimately, EIX will arrive at the Integrated Information/Knowledge Sharing (Stage 4). At this stage, two separate organizations have formed an alliance and implemented interdependent computer systems, which enable them to interactively share single-source information and knowledge whenever it is needed. If engineered correctly, this results in an effective collaborative Information/Knowledge system.

ENTERPRISE INFORMATION ENGINEERING

To achieve Stage 4 EIX, the enterprise's information requirements and architecture need to be systematically identified, engineered, and structured as a baseline for systems implementation and integration. Completing the enterprise's information engineering can be time-consuming and expensive. Consequently, it is recommended that a good understanding of information engineering and implementation successes and failures be accomplished. A process and plan, tailored to fit the organization, can then be constructed to achieve management support. Successful implementation processes vary from one large enterprise-wide information engineering effort to many small segments, each aligned to one ultimate vision and set of objectives. Whichever process is chosen, it must fit the organization's ability to undertake it. The most often used Information Engineering processes involve either Martin (1990) or IDEF (Multiple). This Enterprise Information Engineering activity is one of the most critical elements in long-term EIX success. The references listed within this book and in the Bibliography will provide additional detail implementation assistance.

SUMMARY

The definitions and framework provided provide a basic understanding of EIX terminology. Since EIX is evolving, different authors will expand and revise these basics as appropriate.

Table 1.
EIX Evolutionary Stage

Stage 1	Stage 2	Stage 3	Stage 4
Paper/ Intellectual Package Exchange	Electronic Transaction/ Package Exchange	Interfaced Data Exchange	Integrated Information/ Knowledge Sharing
Appropriate Paper internally transferred; Minimal external transfer	Large amounts of paper internally exchanged; Some high volume specific information is internally & externally exchanged (EDI, IGES, PDES)	Medium amounts of information electronically exchanged directly between independent information generators and consumers through interfaces (CALS)	Large amounts of information and knowledge shared interactively via electronic interdependent integrated systems whenever needed.
Baseline	Efficiency Improvement	Efficiency Improvement	Effectiveness Improvement

A Roadmap for Enterprise Integration

RICHARD J. MAYER
Knowledge Based Systems, Inc. and
Texas A&M University

CAPTAIN MICHAEL PAINTER
Wright Patterson AFB

ENTERPRISE INTEGRATION

The term "integration" has been used since the 1940s to modify the description of a system. The early focus was upon manufacturing systems and integration became the quality metric for determining the level at which the manufacturing system supported the flow of work. Integration efforts were similar to time and motion studies used to improve the efficiency of the production work force. Industrial engineers "methodized" system processes in sufficient detail so that expertise could be duplicated.

By the late sixties, integration began to assume on an even broader definition and the next logical step to methodization was taken. Specifically, we wanted to control the system in such a way that schedule and resource requirements were predictable. Hence, we saw the advent of the Manufacturing Resource Planning (MRP) System, and integration began to connote emphasis on control and prediction.

This level of integration is achievable only in that order: Before you can control the system, you must understand its existing processes, and before you can predict its performance, you must be able to control how the system operates.

By the mid-1970s, integration began to encompass the organization of information resources to support the flow of work. An Integrated Information System was defined as having the ability to share common data and respond gracefully to physical changes which distinguished it from other information systems.

The ability to share common data implies a heavy emphasis on central control and real-time change propagation. This emphasis has become particularly important as technological advances have made it possible to access and manipulate information resident on machines distributed through the enterprise. Simple access has become secondary to the need for consistent control between redundant copies of the same classes of information. More precisely, the control connotations imply that all logical or desired changes to any piece of information should be realized automatically in real-time.

The second quality of evolving physically is motivated by the desire to avoid disruption to logical system functionality as physical components are replaced with new ones. For example, we may want to replace a magnetic disk with an optical disk storage device or one application program for another. Likewise, a retiring company expert may need to be replaced with a new employee.

We can further modify the phrase Integrated Information System to distinguish those systems exhibiting the timely ability to evolve logically and physically. Logical evolution demands the capability to change the information requirements based on changing needs. It may also necessitate enlarging or narrowing the scope of the information system.

Physical and logical evolvability can be called reusability of parts, i.e., to separately manage and control the logical system from the physical implementation in order to isolate the impacts of change. To emphasize the need for information systems capable of evolving physically and logically, the term "Evolving Integrated Information System" (EIIS) is used.

Integration and the Information System

The term "information system" describes systems comprising both automated and nonautomated components responsible for capturing, managing, and controlling information resources. The "informal" system is often far more robust than the "formal" system, thus, an information system might include application programs, the telephone, and your peers. This view of an information system is consistent with current emphasis toward "enterprise" integration as a logical extension of traditional information systems.

There is a common misperception that simply installing an Integrated Information System results in integrating the enterprise. Note, however, that the term integrated modifies information system. This means that the qualities alluded to pertain only to the information system, not to the enterprise as a whole. The information system may serve to integrate or support the efficient flow of work in only a limited portion of the enterprise. Hence, the system may or may not be used as the integration mechanism for the enterprise.

This introduces a less commonly used term, ''Information-Integrated System'' in which the primary mechanism for achieving system integration *is* information. A system is said to be information-integrated when it maintains and uses information about itself to support the flow of work.

It is important to note that an enterprise may include several information systems, each devoted to specific information storage and retrieval tasks. Each information system may represent only part of the overall system, namely the enterprise. The enterprise system is said to be information-integrated when it uses information about itself to integrate all the functions that occur within the enterprise.

This is a subtle, but very important distinction. Integrated information system refers to the qualities of an information system, whereas information-integrated system refers to the relationship maintained between the information system and the environment within which it resides (see Figure 1). This distinction does not imply that physical relations between components of the system become unimportant. However, in an information-integrated system, they do become secondary to information relationships. It is this change of view, towards information-enabled integration, that must come about in order for enterprise integration to succeed.

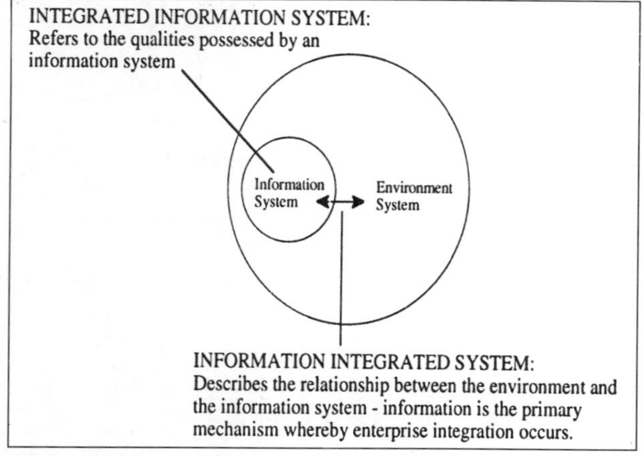

Figure 1. *Integrated Information System versus Information-Integrated System.*

The development capability of an information-integrated system significantly minimizes disruption due to change reduces construction and maintenance costs by leveraging information rather than precision machining each system element.

Once an enterprise begins to evolve toward an information-integrated system, the critical groundwork for many other needed improvements is established. That is to say, initiatives such as Total Quality Management (TQM) and Concurrent Engineering (CE) can be attacked as a natural part of the evolution to the information-integrated enterprise.[1]

INTRODUCTION TO FRAMEWORKS

Stiff competition, computer and technological advancements, increasingly rapid introduction of new products, and other related factors place ever-increasing demands on today's enterprises. To survive, organizations must be more flexible, responsive, customer and quality-oriented.

To meet these demands, a business must understand how to flexibly organize the thinking, planning and execution of its activities.

The term ''framework'' in its most general sense refers to a basic structure, arrangement or system; a ''framework'' serves to hold the parts together.[2]

For example, in physical domains, the airframe components of an aircraft can be referred to as the framework for that aircraft; likewise, the steel supporting structure of an office building can be referred to as the framework for that building. When the parts to be held together are conceptual in nature, then the notion of a framework generally shifts to one of an *organizing structure* for those parts.

One of the key organizing effects of any organizing structure is the provision for (or summarization of) expressions of characteristics of the conceptual parts and the relationships between the parts. Thus, framework is often used as a synonym for the ''conceptual system.'' Further, framework does not conform to a general rule, and only has meaning with respect to the parts that participate in the structure, arrangement, or system.

A framework is a structure for organizing available knowledge about:

1. What situations (e.g., analysis, decision, and design) occur in the evolution of an existing system into an information-integrated system;
2. What methods and tools are available;

3. Under what circumstances particular methods and tools can/should be used in support of a particular development situation;
4. What sequence of application of the rules is the most effective for a particular organization.

Framework refers to an organization of characterized situation types that are known to occur commonly during a system's life cycle. Each characterization, identifies the roles of the personnel in the organization involved in that situation type and the information that must be discovered, decided upon or managed.

For example, a group of system analysts may need to decide how to implement data in an information system, or an analyst may be required to present to the business owners the type of data needed for their company's information system.

With a set of characterizations in mind, we can identify methods and tools that assist in critical information management. We also can identify the circumstances in which information needs to be shared from one development situation to another. In practice, one of the roles of a framework is to help the user select the most effective tool to guide the system developers from the concept of a solution to the reality of implementation, which may be either a new system or a modification to an existing system.

Framework must be differentiated from the term "architecture," which refers to a unifying or coherent form or structure.[2] In the physical domain, it is clear that the framework contributes to the form and the form places requirements on the framework. In the conceptual domain the distinction is not quite as clear. The form, structure, and organization associated with both framework and architecture causes confusion between the two, particularly when describing more specialized concepts. Architectures are important to the enterprise integration effort because on the road from the "as is" situation to the information-integrated enterprise, it is necessary to choose an architectural approach for the information system architecture that will form the backbone of the enterprise. Later in this paper we will present several emerging standards for such information system architectures and discuss how the choice of such an architecture affects the enterprise integration roadmap.

FRAMEWORK COMPONENT OVERVIEW OF THE ROADMAP

Framework, again, refers to an organized representation of characterized situation types that occur during a system's life cycle. A framework can be considered as an organizing structure for classifying information important to the evolution of an organization's information system.

In this section we will focus on the Development Situation Classification Framework (DSCF) of the roadmap for enterprise integration. This framework component of the roadmap is used to put structure into an organization's evolutionary development process.

DSCF frameworks have typically taken the form of a two-dimensional matrix (see Figure 2). Each cell in the matrix represents a class of development situations that share similar participants, viewpoints, and objects of interest.

A development situation characterization should identify: the role of the organization's personnel involved in each situation; the information that must be discovered, decided upon or managed, and a description of the process or method carrying out the activities.

A framework is an architectural representation method that defines the boundaries (and situations) among methods and tools assisting in the design of an integrated system. Frameworks normally come in a generic form (such as the one displayed in Figure 2). One of the steps to setting up a roadmap for enterprise integration at a site is the tuning of the generic framework to a site-specific framework.

Zachman's Framework

One of the most notable generic DSCF frameworks was developed by John Zachman at IBM (see Figure 3). Zachman's pioneering work, reporting on the results of the study of "What, In Fact, is (an) Information Systems Architecture," was based on research in the practice of architects and engineers.

In Zachman's original paper, framework is used as an organizing structure for many information system architectural representations.[3] Zachman presented an overview of the structure of various representations of information systems architectures. His framework attempts to characterize the situations that create or require these representations. The characterizations include identification of the roles, responsibilities, conditions, prior commitments and information required in a situation that results in a need for a particular class of representations.

Thus, one could characterize the resulting descriptive framework documented in Zachman[3] as a morphology. The framework topology is a matrix in which the five rows represent different perspectives and the three columns represent foci of descriptions of

	DATA	USER	FUNCTION	NETWORK
OBJECTIVES/ SCOPE	List of Things Important to the business **BSP** **IDEF5** **IDEF0** ENTITY = Class of Business Thing	List of Scenarios the User Performs **IDEF0** **BSP**	List of Processes the Business performs **IDEF0** **CSF** Process = Class of Business Activity	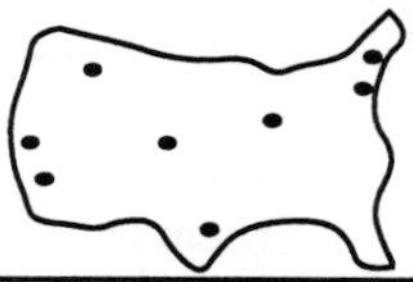
DOMAIN MODEL	e.g. Concept Model **IDEF5** ENT = Bus. Con. Reln = Association	e.g. User Role Description **IDEF3** **IDEF5**	e.g. Business Process Descrip. **IDEF3** **IDEF5**	?
MODEL OF THE BUSINESS	e.g., Entity / Relation Diagram **IDEF1** ENT = Info. Entity Reln = Bus. Rule	e.g. Organization Process Descrip. **IDEF3**	e.g., Function Flow Diagram **IDEF0**	e.g., Logistics Network Node= Bus. Unit Link= Bus. Relatn
MODEL OF THE INFORMATION SYSTEM **IDEF6**	e.g., Data Model **IDEF1x** **ER** ENT = Data Entity Reln = Data Reln	e.g., Transaction model **IDEF3**	e.g., Data Flow Diagram **DFD**	e.g., Distributed System Arch ? Node=I/S Func. Link=Line Char.
TECHNOLOGY MODEL **IDEF6**	e.g., Data Design **IDEF4** **ER** ENT = Segment Reln = Pointer	e.g., Object Design **IDEF4** **Booch**	e.g., Structure Chart **SCG** **IDEF4**	e.g., System Arch ? Node=Hardware Link=Linė Spec.
DETAILED REPRESEN-TATIONS	e.g., Data Design Description ENT = Field Reln = Address	e.g., User Inter-Face Code	e.g., Program	e.g., Network Architecture
FUNCTIONING SYSTEM	e.g., Data	e.g., Scenario	e.g., Function	e.g., Communication

Figure 2. Generic Enterprise Integration Framework.

an information system architecture (see Figure 3). This framework will be described in greater detail later in this report.

Other frameworks derived from the original Zachman framework[4] are organized in a similar way (see Figure 4). The columns represent different areas of focus of a system description (e.g., "Business Objectives (WHY)," "Organization (WHO)," etc.).

In Figure 4, descriptions of the rationale for an information system that supports the business system are represented by the cells in the Objective (WHY) column. The framework's cells can represent other descriptions of objects of interest to the business in the same manner.

The rows represent the various viewpoints within the organization. This dimension organizes the descriptions of the system architecture with respect to multiple viewpoints (e.g., the executive (Concept Description), the manager (Business Description), the programmer (Detailed Description)). Thus, each cell in the matrix represents an enterprise situation and its corresponding system descriptions constructed from the perspective of a particular user.

Uses of a DSCF for Enterprise Integration

A DSCF provides information about the types of situations involved in the system development process. Reported uses of a DSCF at a site include:

1. Communicating the "big picture" of the System Development Process.
2. Providing a "quick roadmap" of the System Development Process for participants.
3. Identifying standard methods and tools use.
4. Identifying applicable tools and methods.
5. Supporting the planning and scheduling of the system development process.
6. Orchestrating the use of integrated tools and methods.
7. Providing a summarized description of the standard development process.

This DSCF framework allows the flexibility required to properly define the development process for a specific project, and provides carry-over of the experience base from project to project within an organization. The installed framework provides the necessary control over the project system development to ensure development success, and provides consistency between projects required to allow multiple project coordination, management consistency, and personnel portability.

Once a framework has been defined, the opportunities for use of this knowledge base include:

1. Context-Defined Tasking.
2. Life Cycle Data Management and Control.
3. Automated Project Status Reporting.
4. Documentation Generation.
5. Automatic Problem Notification.

These capabilities are possible because the framework completely defines the activities that will occur during the development process. The framework also defines the relationships between activities, the objects (documents, code modules, etc.) that will be manipulated during a particular activity, and the roles of people who will be involved in the activity.

Development Situations

Each cell in a framework represents a characterization of an organization's *recurring development situation(s)*. For example, the situation in Figure 5 illustrates a group of corporate managers working with business planners and information systems managers to identify such items as "critical success factors." This cell does not characterize all situations in which corporate management would identify critical success factors, but accounts only for those situations associated with a system development or modification activity.

The individuals involved in the situation, the purpose or context of the occurrence of a situation, and the items/objects that appear, are used, or are produced in that situation play an important part in characterizing a situation.

An example of an object that might appear in a situation would be the corporate mission statement. This object could be used to derive business unit objectives which then might be used to identify specific critical success factors. Thus, understanding the process of discovery, analysis, formulation, or decision-making that unfolds during the situation is important to characterizing the development situation. A description of the process (or processes) that occurs during an instance of that development situation type is a key component of the situation description.

If we examine the object or item description component of a situation type description, we see that the identification of these objects includes the identification of both the *properties* of and the *relations* encompassing the objects.

Figure 9 illustrates the content of a typical cell definition in a DSCF framework with associated inter-cell relations. In the example illustrated by Figure 5, an

	DATA	FUNCTION	NETWORK
OBJECTIVES/ SCOPE	List of Things Important to the business ENTITY = Class of Business Thing	List of Processes the Business performs Process = Class of Business Activity	List of Locations in which the business operates
MODEL OF THE BUSINESS	e.g., Entity / Relation Diagram ENT = Info. Entity ReIn = Bus. Rule	e.g., Function Flow Diagram	e.g., Logistics Network Node= Bus. Unit Link= Bus. Relatn
MODEL OF THE INFORMATION SYSTEM	e.g., Data Model ENT = Data Entity ReIn = Data ReIn	e.g., Data Flow Diagram	e.g., Distributed System Arch Node=I/S Func. Link=Line Char.
TECHNOLOGY MODEL	e.g., Data Design ENT = Segment ReIn = Pointer	e.g., Structure Chart	e.g., System Arch Node=Hardware Link=Line Spec.
DETAILED REPRESEN-TATIONS	e.g., Data Design Description ENT = Field ReIn = Address	e.g., Program	e.g., Network Architecture
FUNCTIONING SYSTEM	e.g., Data	e.g., Function	Communication

Figure 3. *Zachman's Original Framework.*

FOCUS →

PERSPECTIVE ↓

	MANAGERIAL & HUMAN SUBSYSTEM			TECHNOLOGICAL SUBSYSTEM		
	BUSINESS OBJECTIVES "WHY"	ORGANIZATION "WHO"	LIFECYCLE "WHEN"	FUNCTION "HOW"	DATA "WHAT"	NETWORK "WHERE"
CONCEPT DESCRIPTION	List of Business Strategies Objective = 'Acquire Business', etc.	Lisr of Organizational Units Org = High Level Organizational Unit	Business Life Cycle $ T P1 P2 Aggregate of Products	List of Processes the Business performs Process = Class of Business Process	List of things important to the business Entity= Class of Business Thing	List of locations in which the business operates Node= Business Location
BUSINESS DESCRIPTION	e.g. Business Objectives To Meet Strategies	e.g. Organization Chart Structure to control & direct tech & people	Product(s) Life Cycle $ Definition of Product Life Cycle	e.g. Function Flow Diagram Process = Business process I/O = Business resources	e.g. Entity Relationship Diagram Entity= Business Entity Relationship = Business rule	e.g WATS MCI Node = Business Unit Link = Business Relationship
SYSTEM DESCRIPTION	e.g. Definition of measurements	e.g. Organization Structure -Job roles & relationships -Communications & Information	e.g. Process Cycle $	e.g. Decomposition of flow diagram	e.g. data model entity = data entity relationship = data relationship	node = I/S function (processor, storage access etc.) link = line characteristics
TECHNOLOGY CONSTRAINED DESCRIPTION	e.g. Definition of collection methods	e.g. work group -willingness to accept change -work group factors	e.g. Definition of life cycle span	e.g. further decomposition of flow models	e.g. data design entity - segment/row relationship = pointer/key	e.g. system architecture node = hardware/system software line = line specifications
DETAIL DESCRIPTION	e.g. Measures	e.g. Work Group -job satisfaction -motivation	e.g. Specific time frame description -events occuring -issues/concerns	e.g. the Basic Work Unit the Logical Description ("the primitive")	e.g. Database Description entity = fields relation = addresses	e.g. the Basic Work Unit the Physical Description
ACTUAL SYSTEM	e.g. Goals	e.g. People	e.g. Time	e.g. Function	e.g. Data	e.g. Communications

Figure 4. Development Situation Classification Framework (DSCF).

executive's view focusing on the organization goals that the information system must support is chosen. In this cell, the relations/properties of interest specify rules for how the definitions of this level must correspond to those which must be produced in the lower level.

Implications of Cellular Interactions

The description of a development situation must include the description of relationships between differing situations. That is, the "context of occurrence" is an important part of the situation description. To interpret a particular situation, we need to know its context of occurrence. It is this context that allows a particular development situation to have meaning relative to a system development process.

In studies of the allocation of methods to the cells in a DSCF evolved another interesting characteristic concerning framework representations: the form of the representations is structured to allow for easy checking "across representations" of the key decisions. This correspondence-checking is best illustrated in the building architecture domain. Figure 6 illustrates this correspondence between the initial bubble diagrams that are used to decide on the critical connectivity relations between the major functional areas of a building and the space layout diagrams that are used for the initial space allocation decisions. With the representation chosen for the latter, the connectivity constraints of the former can be easily checked.

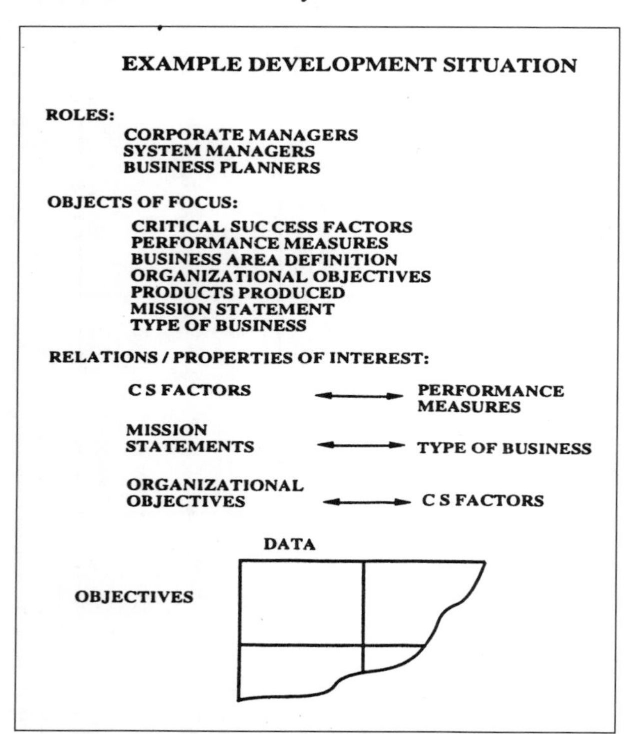

Figure 5. Elements of a Framework Cell.

This concept of interlocking representations carries over into the system development domain. Figure 7 illustrates some of the correspondences across system development situations that need to be supported by the methods characterized within a DSCF for enterprise integration. These correspondences are integration requirements of the methods called for in the framework cells. Integration requirements are not universal, that is, they vary from site to site depending on the needs and rules governing system development. Thus, in frameworks of this type, the rule base associated with the framework (see Figure 8) will actually define the specifics of the type of method and tool integration that are meaningful at a site.

Organizing Enterprise Integration Knowledge with a DSCF

General frameworks, such as Zachman[3], IDEF Users Group[4], and Mayer[5], are useful as reference documents. However, for purposes of application to enterprise integration, the cell descriptions of these reference frameworks will need to be specialized for a particular organization (site-specific).

A site-specific frameworks (SSF) can be thought of as an enterprise knowledge base of system development expertise. In general, an SSF would add the following information to the situation description associated with each cell in a generic framework.

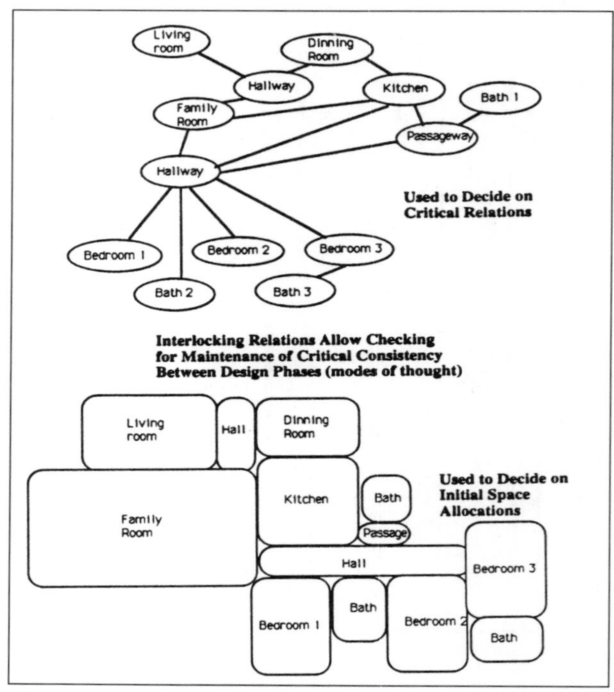

Figure 6. Interlocking Hierarchy of Architecture Representations.

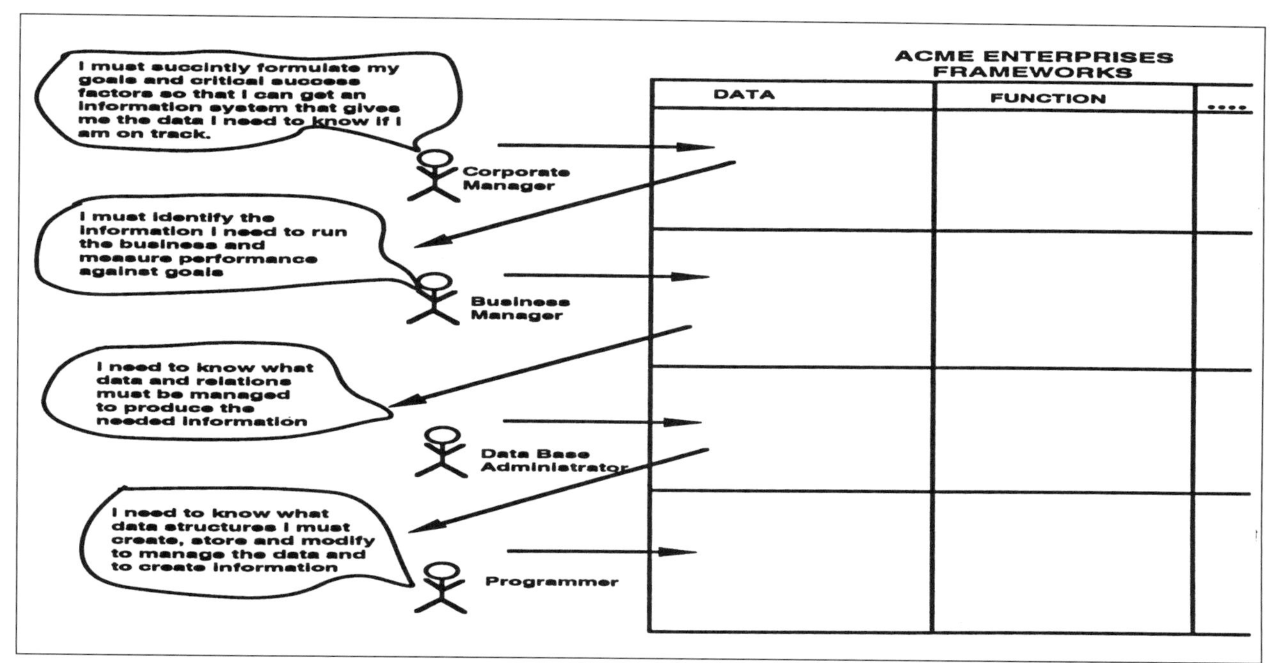

Figure 7. *Hierarchy of Interlocking Methods.*

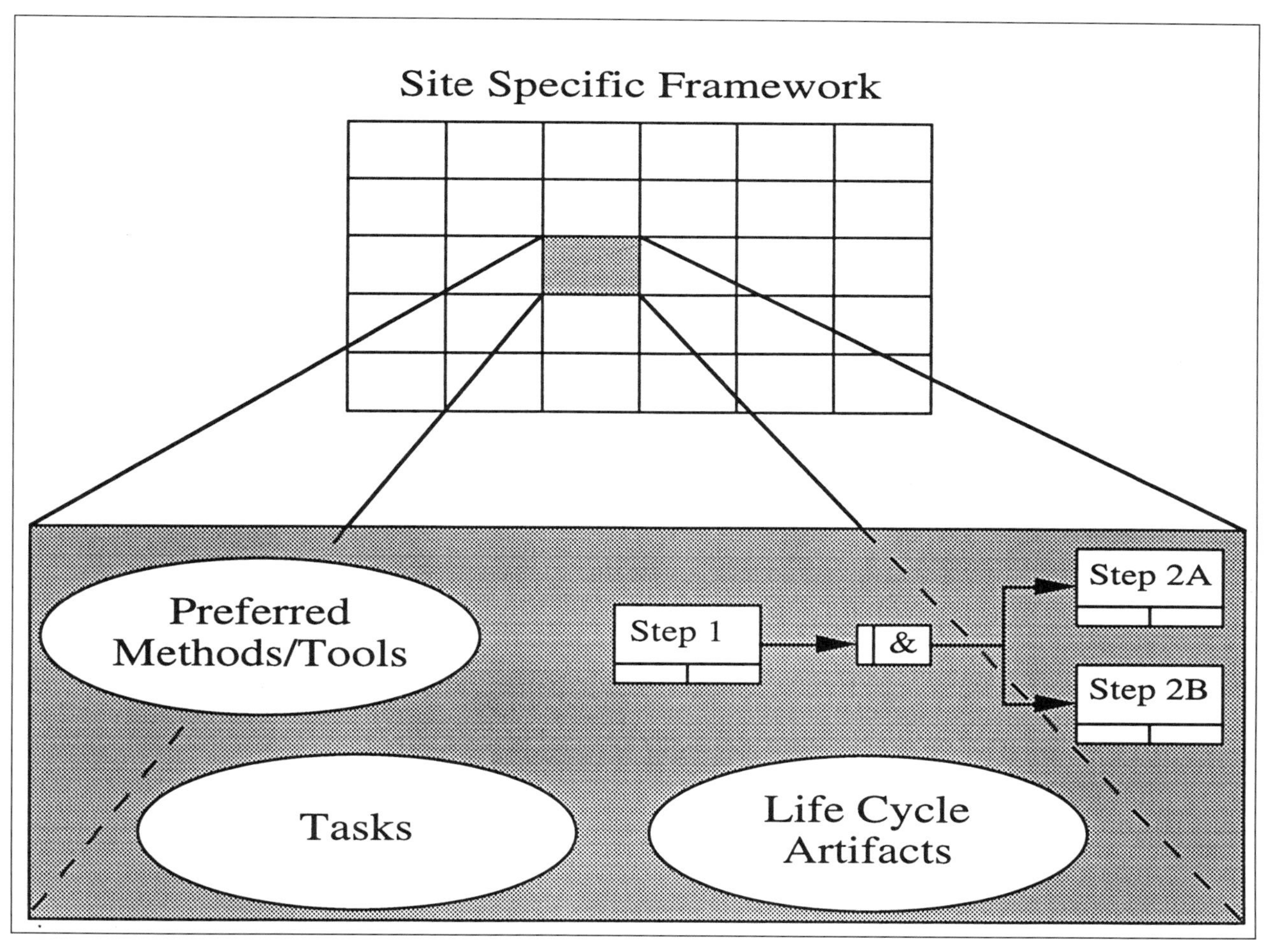

Figure 8. *Site-Specific Framework.*

1. The purpose for the situation type. Objects/ Relations of interest.
2. Information that must be discovered or maintained.
3. Critical decisions that must be made in that situation.
4. Methods, rules and tools recommended for use in the situation, including syntax of the method language, discipline for application of the method, and procedure for use of the method results.
5. Personnel roles involved.
6. Artifacts produced.

The resulting SSF provides a cohesive and consistent architecture within an organization's system development effort.

Note the difference between a framework chart (or graph) and the actual knowledge base that it represents. Figures 2, 3, and 4 display what we think of as a framework (i.e., the matrix). The matrix is actually only a chart that serves as an index to information (situation descriptions) and knowledge (rules, procedures and methods) represented by each cell of the framework chart. All this information constitutes the SSF. The framework chart serves as an organizing structure for the information required to control and manage the system development process.

A bill of material view of the components of an SSF for enterprise integration is displayed in Figure 9. The framework graph contains an illustration for each of the individual cells that indicates the type of questions that the artifacts (produced by the system development activities associated with the situation represented by the cell) should be able to answer. The rules of this management framework appear in the detailed figure of the leaf components. Process models for the overall development activity description and for the activity descriptions of the various roles are contained in the development process description component.

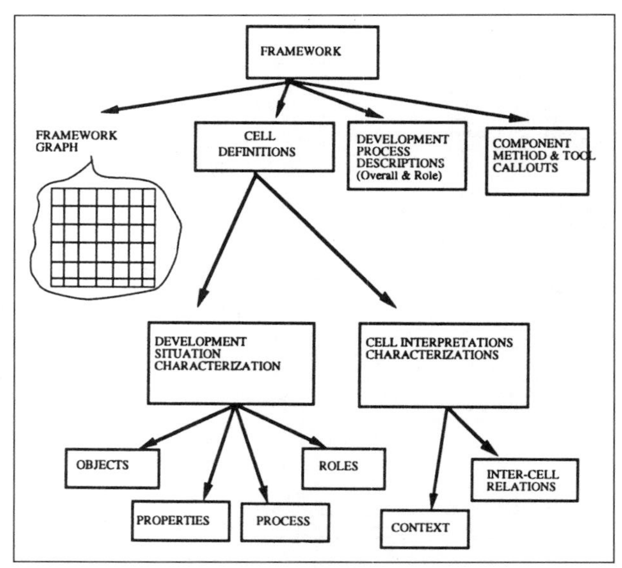

Figure 9. *The Framework of BOM.*

Formulating Site-Specific Frameworks

A technique to aid in the definition of SSFs is based on the questions that should be answered by that situation type. We have found that rather than trying to initially identify the situations, roles, objects, and relations of interest, it is often easier to collect the questions that we would like to answer from the artifacts produced by the (yet unidentified) situation types. This approach is consistent with Zachman's initial intuition that the cells represent descriptions of the actual system that reduce risk by explicitly documenting the various decisions and views of the system. That is, a particular piece of data should exist in the system to directly support a business performance or operational information requirement.

In a typical framework, we can classify general question types, and within each question we can identify general question templates. One such question type is referred to as an introspective question. Introspective questions are associated with cells that are directed at the personnel who have the perspective indicated in the row label of the cell row. "What are the goals of the enterprise that are affected by the system?" would be associated with the cell in row one and column one of the Zachman framework (see Figure 3). This question is also a good example of a recurring question schema.

One of the general templates for introspective questions is <what, who, where, when> <be verb form> the <column focus> of the <row perspective> of the <system name>? Part of our ongoing work with frameworks for enterprise integration is focused on the development of an expert system for the generation of site-specific frameworks using this question template concept. Even without an expert system, use of these templates can accelerate the definition of a SSF.[5]

Site-Specific Situation Defined

Once the question templates have been specialized for a site, the next step in the SSF cell formulation is the definition of the situations in which questions could be answered. The roles for a situation definition will include personnel responsible for providing the information to answer a question. The objects of interest often represent the elements of the answer to the question, as do the object relations.

Once the situation descriptions have been formulated, the next step is to specify the rules that govern the question-answering process. These rules can reference the results of other cells as well as the objects and roles of a particular cell. The process definition itself takes the form of a set of process diagrams. It is also at this stage that the selection of methods (and tools) to support this process application (and rule enforcement) occurs. Finally, the rules for how the answers to the questions are to be used and managed must be defined. This process can be accelerated by having access to generic or reference frameworks.

INTRODUCTION TO METHODS

A *method* is an organized, single-purpose discipline or practice.[7] A method may have a formal theoretic foundation; however, most do not. Generally, methods evolve as a distillation of best practice experience. The term methodology has at least two common usages. The first refers to a class of similar methods; for instance, one may hear reference to the *function modeling methodology*, meaning methods such as IDEF0 and LDFD.* In another sense, methodology is used to refer to ''a collection of methods and tools, the use of which is governed by a process superimposed on the whole'' (Coleman 1989). Thus, the criticism that a tool (or method) has no underlying methodology is common. Such a criticism is often leveled at a tool (or method) which has a graphical language, but for which no procedure for the appropriate application of the language or use of the resulting models is provided. By ''tools,'' we mean a software system designed to support the application of a method.

Although a method may be informally thought of as simply a procedure for doing something (as well as, perhaps, a representational notation), it is more formally described as consisting of three components, as illustrated in Figure 10. Each method has a definition, a discipline, and many uses. The definition contains the concepts involved and the theory behind the method. The discipline includes the syntax of the method and the procedure by which the method is applied. Many methods have multiple syntaxes which have either evolved or are used for different modeling aspects. One of the most visible components of a method is the language associated with the discipline.

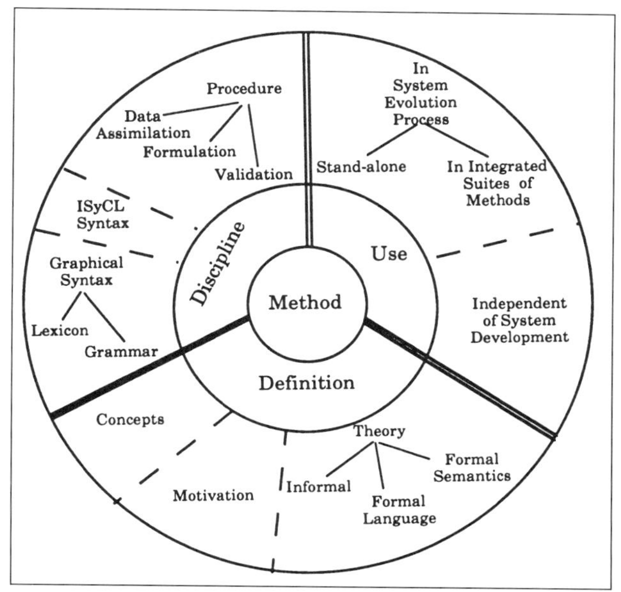

Figure 10. *Anatomy of a Method.*

Many system analyses and engineering methods use a graphical syntax to provide visualization of data in such a way that key information can be easily extracted.

The primary purpose of a method is to ensure good practice experience in order to raise the performance of novice practitioners to an expert level. Graphical facilities provided by method language serve, not only to document the analysis or design process undertaken, but more importantly, to highlight important decisions or relationships that must be considered during method application.

In traditional architecture a bubble chart is used to collect the customer's needs. These diagrams are comprised of circles, or bubbles, representing living spaces and simple arcs connecting the bubbles. The arcs represent relationships to be maintained between individual living spaces. Where these relationships are particularly strong, the architect notes the strength of the relationship by using a thick line for the arc. Space is allocated by using a spatial relationship diagram which looks something like a floor plan.

The architect chose to use a bar chart rather than a spatial relationship diagram to allocate the available floor space to individual rooms. By using the spatial relationship diagram, however, the architect is able to rapidly inspect whether or not the constraints identified in the bubble charts are maintained in the design. In fact, the use of spatial relationship diagrams rather than a simple bar chart exposes possible realizations of

**IDEFO is an activity modeling method[6] derived from the Structured Analysis and Design Technique.[9] LDFD is a Logical Data Flow Diagramming method.*

a design that fit the bubble chart limitations. For those designs that do satisfy the constraints, a qualitative assessment of the suitability of the design can be directly determined.

Creating visualizations provided by methods is of critical use to the practitioner. Visualization allows extraction of additional information from the data provided. Such visualization-derived information, when used in conjunction with knowledge-based tools, transforms information into knowledge.

Today's methods and method support tools are at the point where CAD systems were in the early '60s. Many of today's method support tools amount to little more than expensive pencils. What we would like to have are smart tools that can take full advantage of the power of visualization, a capability CAD systems are now moving toward.

To move in this direction, we must have methods available that are specifically designed to take advantage of both good practice for a given development situation and visualization to highlight important relationships between different perspectives relevant to the system development process. By treating methods as systems in their own right, we should be able to develop the mathematical foundations, engineering tools, and design techniques that will allow the rapid engineering of new methods with predictable effectiveness. Likewise, we should be able to reverse-engineer existing methods to determine the most appropriate means for promoting inter- and intra-method integration. There are different kinds and levels of integration obtainable between methods considering each class or type of method is intended to represent a unique perspective of the system under consideration. There may/may not be any direct correspondence between concepts used by various methods.

There are two kinds of integration that can be achieved between methods. The first is called interoperability, which necessitates taking advantage of procedural similarities, or complements, in the application procedures of a given pair of methods. Within a specific system development process, the identification and exploitation of opportunities to coordinate method application activities can prove beneficial. The second kind of integration is characterized by data sharing, in which opportunities emerge where there is overlap between method concepts.

We assume that there will always be a place for special-purpose methods designed to address systems development on a small scale where integration concerns are secondary. There are numerous examples of "quick and dirty" method designs for short-term, small-scale systems development. This phenomena accounts for literally thousands of ad hoc methods in use today. On a scale representing all possible levels of method integration, special-purpose methods requiring no integration with other methods would be found at the near end of the spectrum (see Figure 11).

Moving to where the need for physical integration exists, a level appears where pairs of methods have some conceptual and/or procedural overlap. In this case, where conceptual overlap exists there is a one-for-one correspondence between syntactic elements in the two method languages. Procedural overlap is accommodated by a presumed sequence of application.

The far end of the spectrum fits logical integration between methods. Here, there is an additional requirement to understand the indirect relations between nonoverlapping concepts among pairs of methods. This understanding enables data sharing at the logical level in addition to empowering the system to predict or infer necessary changes to individual method perspectives. Likewise, procedural integration between methods must be generated according to the system development procedure specified for a particular problem. At this level, the environment uses information about methods as the primary mechanism for both procedural and data integration.

The IDEF Family of Methods for Enterprise Integration

IDEF (Integrated Computer-Aided Manufacturing (ICAM) DEFinition) methods are used to perform modeling activities in support of enterprise integration. The original IDEFs were developed for the purpose of enhancing communication between people who needed to decide how their existing systems were to be integrated. IDEF0 (Function Modeling Method) was designed to allow an expansion of the description of a system's functions through the process of function separation and categorization of the relations between functions (i.e., in terms of Input, Output, Con-

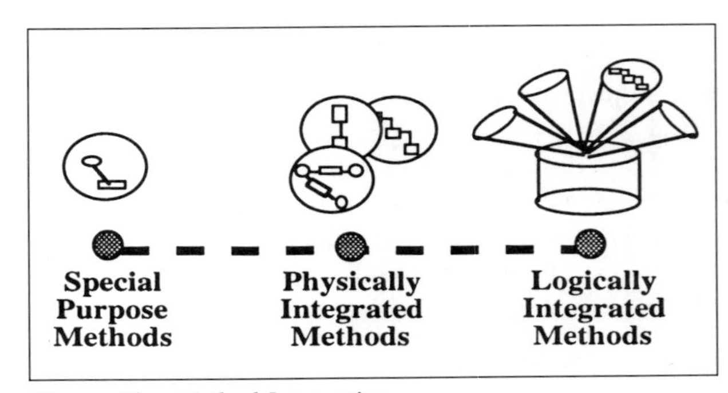

Figure 11. Method Integration.

trol, and Mechanism classification). IDEF1 (Information Modeling Method) was designed to allow the description of the management information that an organization deems important in accomplishing its objectives. IDEF2 (Simulation Modeling Method) was originally intended as a user interface modeling method. However, because the ICAM program needed a simulation modeling tool, the resulting IDEF2 was a method for representing the time-varying behavior of resources in a manufacturing system and provides a framework for defining math-model-based simulations.

The lack of a method which would support the structuring of descriptions of the user view of a system has been a major shortcoming of the IDEF system of languages. From a methodology viewpoint, the basic problem is the need to distinguish between a description of what a system (existing or proposed) is supposed to do and a representative simulation model that will predict what a system will do. The latter was the focus of IDEF2; the former is the focus of IDEF3 (Process Description Capture Method). There are two additional description capture IDEF methods under development. IDEF5 (Ontology Description Capture Method) is a method for knowledge acquisition; IDEF6 (Design Rationale Capture Method) is a method for capture of information system design rationale.

The second class of IDEF developed methods focuses on the design portion of the system development process. That is, the methods ensure the best known method for design with a particular technology (or class of technology.) Currently, there are two IDEF design methods, IDEF1X (Data Modeling Method) and IDEF4 (Object-Oriented Design Method). IDEF1X was developed to assist in the design of semantic data models. IDEF4 was developed to address the need for a design method to assist in the production of quality designs for object-oriented implementations. IDEF4, like IDEF1X, is intended to service the needs of the systems designers and programmer analysts who are building and evolving large information systems. Unlike IDEF1X, which encapsulates a design method for relational database design, IDEF4 encapsulates the principles for design of object-oriented applications and databases. Figure 1 illustrates the planned areas of application for the IDEF methods relative to an updated Zachman framework for information system architectures.[3,4,8] Figure 12 shows the additional IDEF methods that are planned for development over the next two years. These methods will provide a rich complement of method capabilities for enterprise integration efforts.

Suite of IDEF Methods

IDEFØ	**Function Modeling**
IDEF1	**Information Modeling**
IDEF2	**Simulation Modeling**
IDEF1x	**Data Modeling**
IDEF3	**Process Description Capture**
IDEF4	**Object-Oriented Design**
IDEF5	**Ontology Description Capture**
IDEF6	**Design Rationale Capture**
IDEF8	**User Interface Modeling**
IDEF9	**Scenario-driven Info Sys Design Spec**
IDEF10	**Implementation Architecture Modeling**
IDEF11	**Information Artifact Modeling**
IDEF12	**Organization Modeling**
IDEF13	**Three Schema Mapping Design**
IDEF14	**Network Design**

Figure 12. *Next Generation IDEF Methods.*

A ROADMAP FOR ENTERPRISE INTEGRATION

A roadmap for enterprise integration can now be constructed. Such a roadmap consists of: an established set of management goals and policies, a choice of a generic DSCF, the production of a SSF, and the determination of a recommended development path through the SSF. The last element takes the form of a standardized system development process for the enterprise, i.e., a framework. Under this view of framework, the parts being structured are time-sequenced life cycle activities such as analysis, design, implementation, maintenance, or decision-making activities. These types of frameworks are also referred to as System Development Life Cycle Models.[7,8,10] We refer to these frameworks as System Development Process Frameworks (SDF). Figure 13 shows a graphical representation of a generic SDF.

Information System Architectures for Enterprise Integration

An important component of an enterprise integration roadmap is the choice of the information system architecture itself. Often, the architecture is also referred to as a framework. Under this view, the parts being structured are the databases, networks, operating systems, integration utilities, etc. of the system itself. We refer to this type of framework as the platform view. There has been considerable research and development activity in the United States and in Europe to establish standard architectures for large-scale engineering, manufacturing and business information systems. Figures 14 (ISO, IISS, IDS), 15 (DICE), 16 (CIM/OSA), and 17 (DKMS, ASDW, IICE) show examples of the evolution of these Integrated Platform

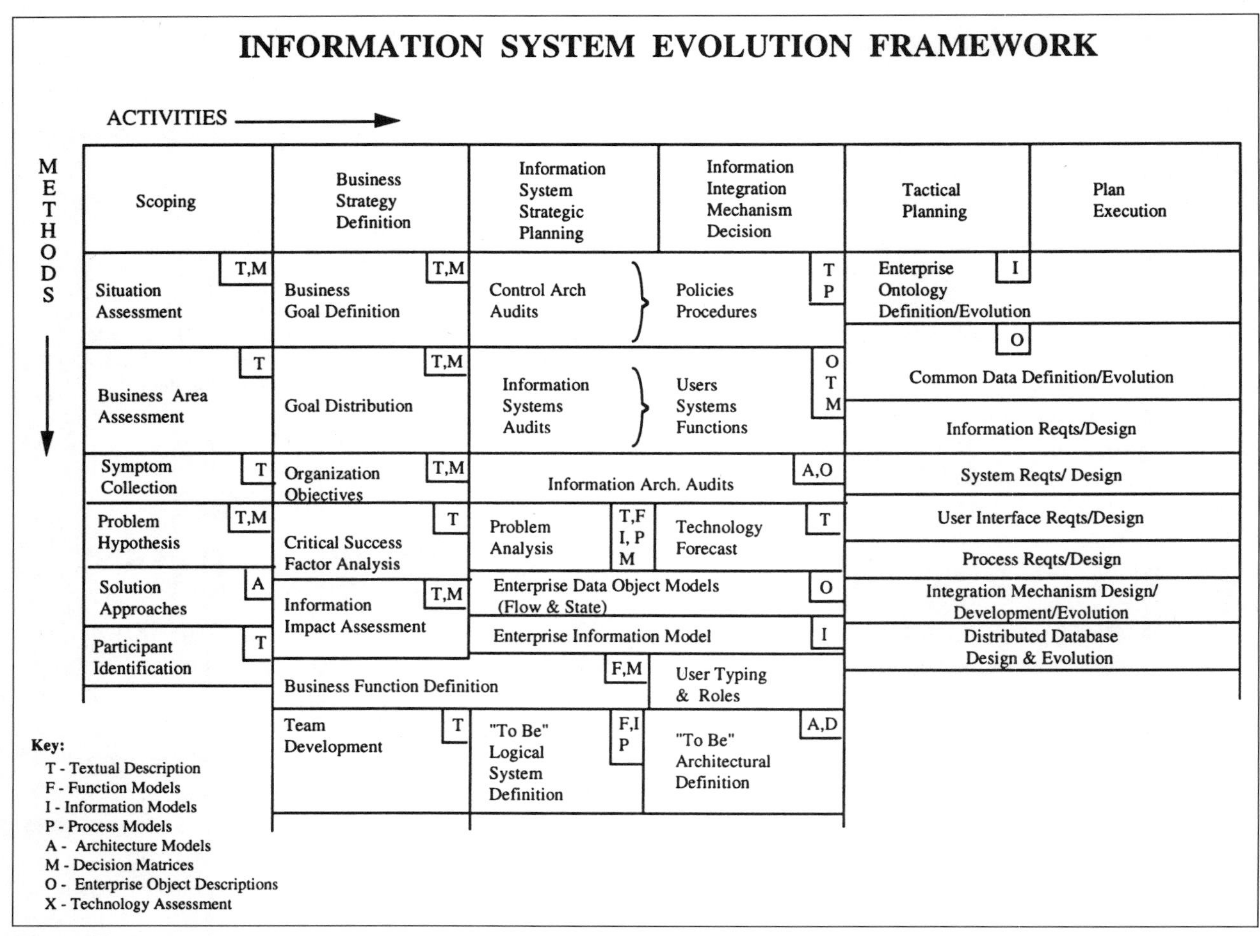

Figure 13. *A Roadmap for Enterprise Integration.*

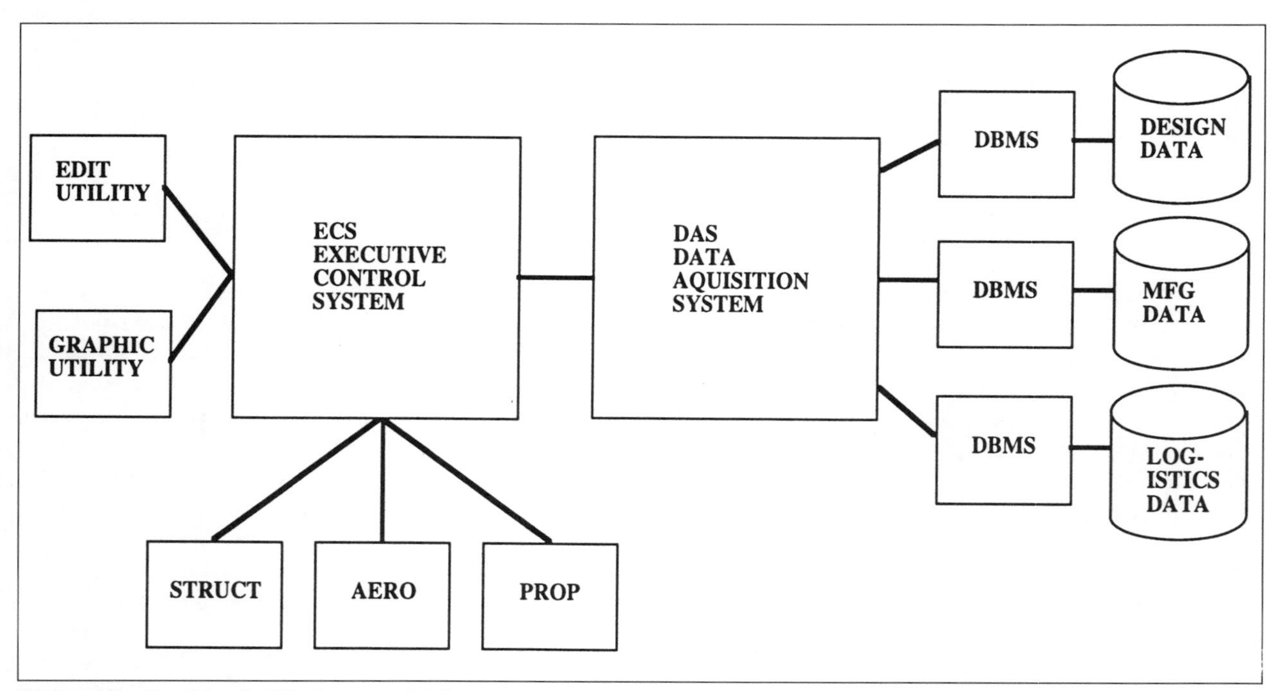

Figure 14. *Traditional ISO Conceptual Schema Based Architectural Framework (circa 1983).*

Frameworks (IPFs). The choice of which strategy to choose is still a difficult question because these standards are just beginning to emerge. It is likely that a particular enterprise will actually end up adopting several strategies for different segments of its operations.

Conclusions

In this paper we have characterized an information-integrated enterprise and have also described the use of Development Situation Classification Frameworks and Site-Specific Frameworks as mechanisms to provide structure through the complex decision processes necessary to predictably evolve toward such enterprise integration. Development Situation Classification Frameworks serve well as the reference from which Site-Specific Frameworks can be created because they enable the appropriate selection of integration methods, tools, and information system architectures needed to successfully implement and evolve toward information-integrated enterprise systems.

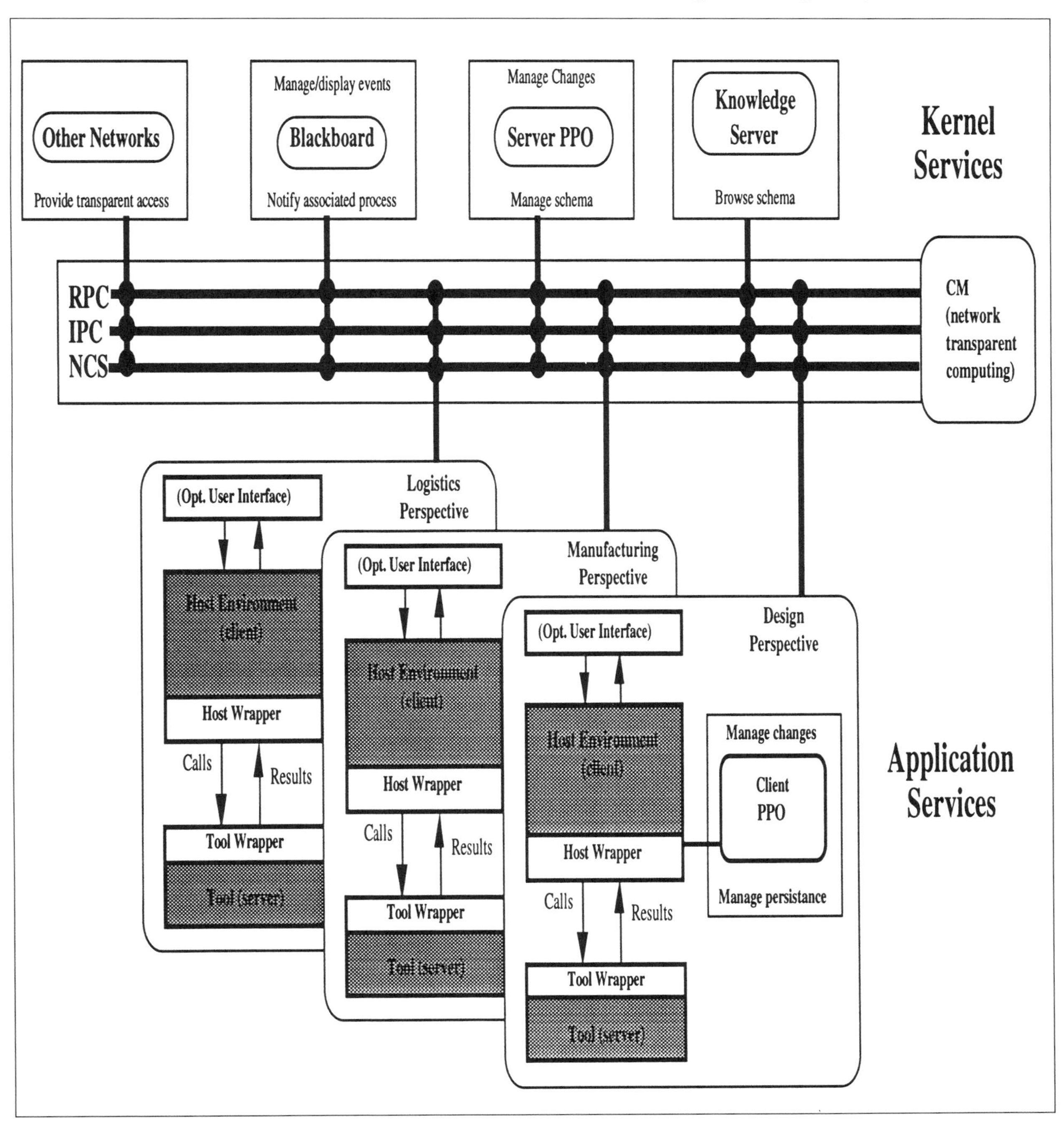

Figure 15. DICE Rose Architecture (circa 1990).

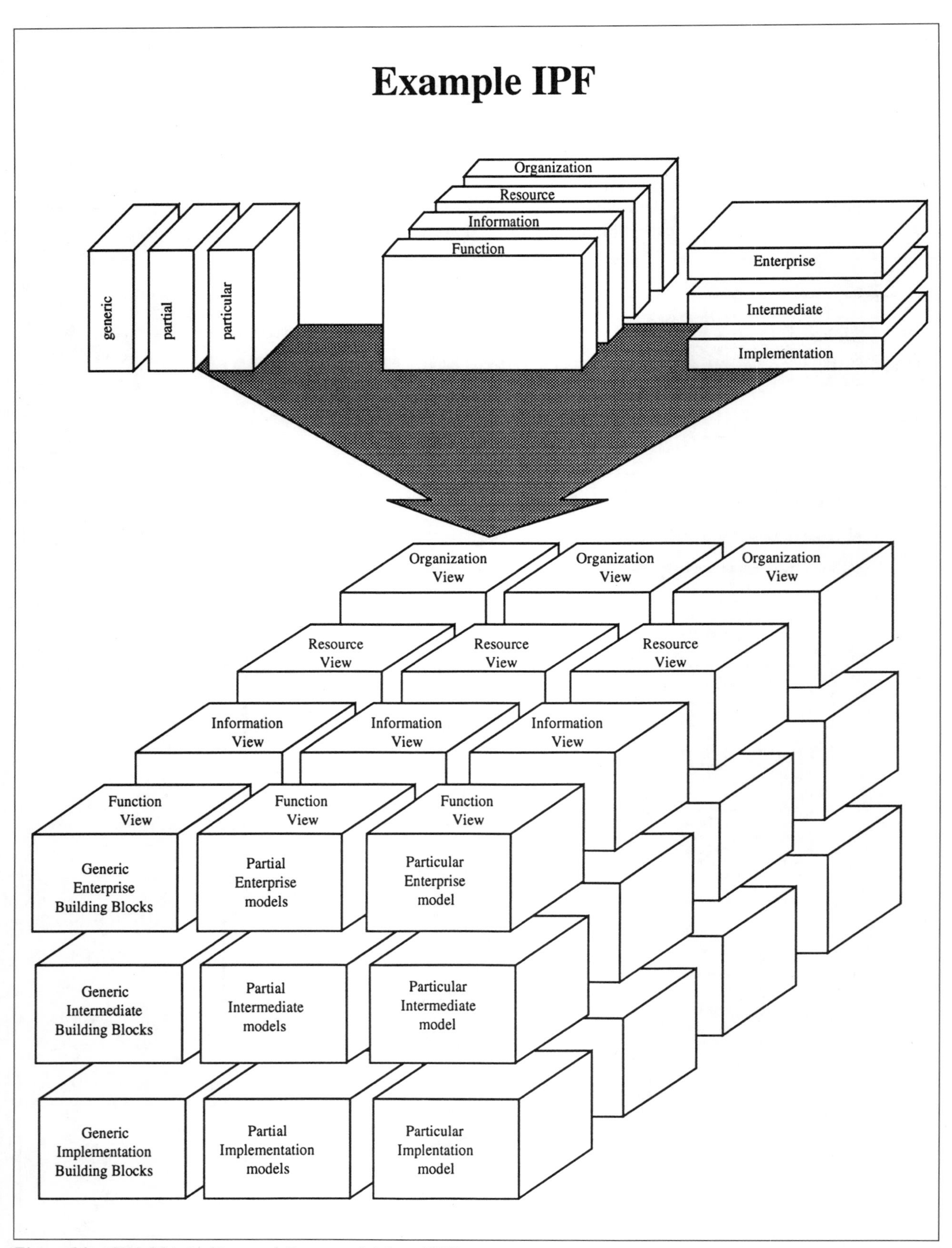

Figure 16. *CIM-OSA Architectural Framework (circa 1989).*

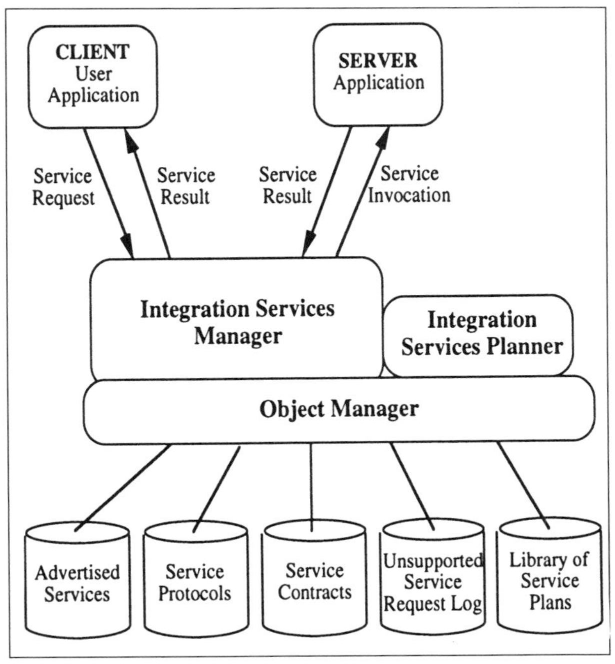

Figure 17. *Integration Services Based Architecture (circa 1991).*

REFERENCES

1. Painter, M. "Information Integration for Concurrent Engineering (IICE): Program Foundations and Philosophy," Conference Proceedings for the IDEF Users Group, May 1991.
2. *The Merriam-Webster Dictionary*, Simon & Schuster, New York, NY, 1986.
3. Zachman J. A. "A Framework for Information Systems Architecture," IBM Los Angeles Scientific Center, G320-2785, March 1986.
4. "Working Group 1 (Frameworks) Technical & Test Committee," IDEF Users Group, *The IDEF Enterprise Framework*, Document IDEF-UG-0001, Ver. 1.0, Jan. 24, 1990.
5. Mayer, R. J. "A Framework Generator," Draft Technical Report, Knowledge Based Systems Laboratory, July, 1990.
6. Mayer, R. J. and Su C. J. "An Intelligent Machining Process Assistant," In revision, *Journal of Intelligence in Manufacturing*.
7. Coleman, D. S. "A Framework for Characterizing the Methods and Tools of an Integrated System Engineering Methodology (ISEM)," Pacific Information Management, Draft 2 Rev. 0, May 1989, p. 2.
8. Mayer, R. J. et al. "Analysis of Methods," Knowledge Based Systems Laboratory Technical Report (KBSL-89-1001), 1989.
9. Ross, Douglas T. "Structured Analysis (SA): A Language for Communicating Ideas," *IEEE Transactions on Software Engineering*, January 1977.
10. *Integrated Design Support System (IDS)* AFHRL-TR-89-6: Volume I–"Executive Overview"; Volume II– "IDS Introduction and Summary"; Volume III–"IDS Requirements"; Volume IV–"IDS Task Results"; Volume V–"IDS Software Documentation." AFHRL, WPAFB, OH, December 1989.

BIBLIOGRAPHY

Concepts and Terminology for the Conceptual Schema and the Information Base, ISO/TC97/SC5/ WG3, 1982.

DARPA Initiative in Concurrent Engineering (DICE): Red Book of Functional Specifications for the DICE Architecture, February 28, 1989. Contract MDA972-88-C-0047, Concurrent Engineering Research Center, West Virginia University.

"A Design Knowledge Management System" (DKMS), SBIR Phase I Final Report, April 1990, Knowledge Based Systems, Inc. Contract F41622-89-C-1018, AFHRL, WPAFB, OH.

"Framework Programmable Platform for Advanced Software Development Workstation: Concept of Operations Document," Knowledge Based Systems, Inc., Prepared for NASA-Johnson Space Center, RICIS Pro gram: Subcontract Number 077: Cooperative Agreement Number NCC 9-16, September, 1990.

Judson, D. L. "Integrated Information Support Systems," 1986; "Integrated Information Support System (IISS): An Evolutionary Approach to Integration," Manufacturing Technology Division, Materials Laboratory, Air Force Wright Aeronautical Laboratories, 1985.

"Open System Architecture for CIM," Research Reports ESPRIT, Project 688, Amice, Volume 1. Springer-Verlag, New York.

System Development Methodology User's Manual, Hughes Aircraft Company, UM170131000, Oct 1983, Vol 7.

Section Three: Industry and Government Standards

Section Three papers are:

- *Implementing Standards for Interdepartmental Document Sharing.*
- *An Overview of Electronic Interchange Standards.*
- *PDES: The Enterprise Data Standard.*
- *Computer-Aided Acquisition and Logistics Support (CALS) Primer.*

This section contains articles describing industry and government data exchange standards, their development, implementation, and use. In the first paper, Ybarra discusses the internal needs for and benefits of standardization through a review of standards implemented at McDonnell Douglas. This process benefited external customers while assisting internal data exchange.

Rohde, in the second paper, describes a variety of standards developed with details of EDIF, IGES, IPIC, and VHDL. He provides a list of advantages and cautions to be considered before system implementation.

The development of the PDES standard is described by Carringer. He provides a brief history, definition of PDES goals, and implementation architecture description. A description of the PDES Incorporated project, an industry-funded cooperation program, provides a good framework for implementation of such standards.

The last paper in this section is a foundation description of the CALS (Computer-Aided Acquisition and Logistics Support) program scope and implementation. Pechersky describes how and why the CALS concept was initiated. The CALS program may significantly reduce information exchange costs. These key industry and government standards enable internal and external information exchange to be accomplished sooner and at costs lower than proprietary customized approaches.

Presented at the CASA/SME AUTOFACT '89 Conference, October 31-November 2, 1989

Implementing Standards for Interdepartmental Document Sharing

DANO YBARRA
OMS, Incorporated

Conversion to standards was successfully accomplished at McDonnell Douglas despite budget constraints, shortages of equipment and human resources, and large physical distances between divisions. While there are still improvements possible to the CALS initiative, the specifications contained therein serve as a valuable framework to create universally compatible hardware and software configurations. Once compatibility is finally achieved within and between companies, the electronic transfer of data will prevail, and the era of the "paperless" office will be realized.

INTRODUCTION

McDonnell Douglas is a large corporation best known for its aircraft designs. Customers include a variety of government agencies and a cross section of government related industries, oil companies, and international governments and corporations. The Department of Defense (DoD) is one of its major customers. Therefore, when the DoD specified a set of standards for the transmission and reception of electronic media, McDonnell Douglas elected to adopt them.

Even though the decision to standardize transmission of electronic data was prompted by DoD requirements, McDonnell Douglas benefited greatly in the process. Standardization enabled McDonnell Douglas to promote internal communications—the sharing of files and transmission of data—within and between departments.

DoD specified adherence to the Computer-Aided Acquisition and Logistics Support (CALS) initiative, specifically the Federal Information Processing Standards Publication (FIPS PUB) 128, 1987 March 15, U.S. Department of Commerce/National Bureau of Standards. Contained in this publication are Computer Graphic Metafile specifications which adopt the ANSI X3.122-1986 Publication, American National Standard for Information Systems and the International Standards Organization (ISO) standards for graphical file storage and transfer.

The Computer Graphics Metafile (CGM) is a computer file that contains device-independent descriptions of computer graphics images or pictures. Since one critical means of communication between DoD and McDonnell Douglas is chiefly of a graphical nature, CGM was the central software standard to which McDonnell Douglas adapted.

THE INTERNAL NEED FOR STANDARDIZATION

The Engineering Technology Division (ETD) within McDonnell Douglas was given the primary task of testing and implementing the standards to comply with DoD contractual commitments. The ETD saw the task as an opportunity to finally coordinate the variety of hardware and software distributed throughout this large, departmentalized company.

McDonnell Douglas had never procured computer products with hardware or software conformity in mind. As long as the hardware or software solved the needs of the moment, it was purchased. "Until last year," said Tom Young of the ETD, "the ability to communicate from one software package to another was not a consideration for purchase." Once a project was complete, the hardware and software assigned to it would be "adapted" to a new project or simply put into storage. Due to this rather carefree policy, McDonnell Douglas had difficulty utilizing its computer resources to their fullest extent.

McDonnell Douglas has invested billions of dollars

in hardware. The computers are grouped in eight clusters, each cluster containing multiple VAXs, IBM mainframes, and Crays, and hundreds of Microvaxes, Suns (and other workstations), personal computers and Macintosh computers. Linked together primarily via Ethernets, these clusters, together, serve over 40,000 users.

Along with this assortment of computers comes an assortment of operating systems. DOS, Quickdraw, UNIX in various forms, VMS, IBM VM and IBM MVS—various versions of each—are all in use. Although effort was exerted to keep each computer running the newest operating system, complications slowed the process. Many times, for instance, current software was incompatible with a new version of an operating system, and massive adaptations would have to take place.

Due to the widespread inconsistencies, departments were incapable of sharing data with other departments, and sharing data within a department was even in question. Not only were there operating system incompatibilities, each of the thousands of software products used by McDonnell Douglas stored data in its own proprietary file format.

Benefits of Standardization

To the DoD, the ability to receive and transmit data in a consistent electronic format is of utmost importance. Massive quantities of information must be distributed, referred to, and tracked in order to run and maintain the highly complex DoD products. Reliance upon paper documents is exceedingly cumbersome, leading to great expense and the repeated loss of vital information through misplacement.

Speaking on August 5, 1988, William H. Taft IV, Deputy Secretary of Defense, commented on the need for adoption of standards to facilitate the electronic transfer of data. "A 9600 ton Navy cruiser carries 26 tons of manuals on how to maintain and operate its complex systems," he said, speaking of the need to eliminate paper. Furthermore, he stated, "If ... [an engineering drawing is] ... misfiled, it's lost for eternity." Thus his recommendation: "Effective immediately, plans for new weapon systems and related major equipment items should include use of the CALS standards."

For McDonnell Douglas, adherence to CALS standards meant accruing a whole host of associated benefits. Hardware and software could easily be adapted from one project to another. Data saved one day could be read in the future without programmers having to spend multiple man-months to write file format converters. Reliability, supportability, and maintainability would be greatly enhanced by uniformity, since changes could be made to all affected parts across the board. In addition, the automatic interchange of technical information and documentation would be made possible, and the massive flow of paper documents replaced by digital file exchanges. Finally, workers would no longer have to learn new software for each new project—saving much time and expense—and data could be shared by all users, workgroups, and departments.

To everyone's benefit, hardware and software vendors faced with widespread conformity to standards began to design their products to support the standards. Purchasing would be simplified for the DoD and McDonnell Douglas since their procurement departments could specify conformity to a well-documented standard. In this way, they could avoid the usual restrictions designed to discourage procurement specifications from favoring proprietary software.

STANDARDIZATION

ANSI and ISO specifications (as contained in the ANSI X3.122-1986 Publication and the ISO DP 9636, 6 December 1986) offer standards not only for data types, but for hardware interfaces as well. Thus, in addition to the CGM, these specifications include the Computer Graphics Interface (CGI). While the CGI portion is in draft form, it serves as a guideline for standardization.

Compliance with these specifications means that heterogeneous host systems are linked by a common network (CGI) and all software generates and interprets a common file format (CGM). CGI bridges the gap between hardware and CGM allows sharing of all files.

CGMs can be displayed and edited by bringing them into a software package that interprets CGMs. These packages can also be used to convert files to and from CGM formats.

The only remaining difficulties lie with areas in which standards are insufficiently defined in all their aspects. For instance, while the CGM standard itself is well-specified, implementation is not addressed. Therefore, not all software vendors use the same implementation, and adjustments must be made to obtain full compatibility. In addition, though a common network interface is available to interconnect all types of hardware, there is yet to emerge a common network protocol. Nevertheless, this problem can be solved through the use of protocol emulators.

IMPLEMENTATION AT MCDONNELL DOUGLAS

The McDonnell Douglas approach to implementing the standard involved multiple hardware and in boards for the Ethernet network, multiple software installs for the network software, and a number of made-to-order translators, emulators, and interpreters to enable software to interpret and generate CGMs.

Obtaining software and hardware that supports the standards is typically the key to successful implementation of the standards. But McDonnell Douglas had millions of dollars invested in software—as well as in hardware and peripherals—that it did not want to discard. Thus, system integrators began by adapting in-house, custom software by adding to it a CGM generator.

The CGM generator enabled McDonnell Douglas' custom software to generate CGM files. In the initial stages of the implementation, these files were used only for data storage and retrieval. Meanwhile, system integrators wrote interpreters that allowed this same custom software to read-in CGMs for the purpose of editing them.

The next step was to acquire printer language information for the printers they had in-house, including QMS Lasergrafix, QMS PostScript, HP Laserjet, plotters, and a bevy of other printers. Using the information they obtained, the system integrators wrote translators capable of accepting CGM files as input and generating files in the format the printer understood.

At the same time, QMS, Inc. designed a high-speed, high-resolution color and monochrome printer that attaches to a network and spools, directly interprets, and prints CGM files. This network print station, because it is available to anyone on the network, eliminated the need for a lower resolution screen capture printer at each workstation.

Thus, with CGM generators, interpreters, and translators in place, all the software could generate and edit CGM files, while all printers could print from them.

With this internal system up and running, the system integrators began testing various software packages running on mainframes, PCs, and Macintoshs to see if their CGM generators and interpreters were compatible with McDonnell Douglas'. Due to the implementation loophole, not all packages were immediately compatible, and collaboration was necessary. System integrators chose software packages from companies that were willing to work with McDonnell Douglas to make both parties' CGM files compatible.

With all of the generating, transmitting, and interpreting of CGMs, one major goal has been achieved at McDonnell Douglas: people are communicating across the vast distances between companies, divisions, and departments, as well as software environments, hardware collections, and peripherals.

CONCLUSION

Conversion to standards was successfully accomplished at McDonnell Douglas despite budget constraints, shortages of equipment and human resources, and large physical distances between divisions. While there are still improvements possible to the CALS initiative, the specifications contained therein serve as a valuable framework to create universally compatible hardware and software configurations. Once compatibility is finally achieved within and between companies, the electronic transfer of data will prevail, and the era of the "paperless office" will be realized.

BIBLIOGRAPHY

Adair, Weldon B. Jr., P. E., "Computer Graphics Integration at the NASA Johnson Space Center," in Proceedings of NCGA Graphics '89 Conference, (August 1988), pp. 337-345.

Arnold, D. B., and Bono, P., CGM and CGI, Springer-Verlag, Berlin Heidelberg, 1988.

Clainos, Deme, "CALS and the DoD: Taking Command - Implications for the Rest of the Industry," from Dataquest Document Processing Conference, (May 18, 1989), Compound Document Processing: Clarifying the Vision and the Opportunities.

Computer Graphics Interfacing Techniques for Dialogues with Graphical Devices (CGI), ISO DP 9636, 6 December 1986; also dpANS X3.161.

Computer Graphics Metafile (CGM), Federal Information Processing Standards Publication, FIPS PUB 128, 16 March 1987.

Computer Graphics Metafile for the Storage and Transfer of Picture Description Information (CGM), ISO 8632-Parts 1 through 4: 1987; also, ANSI/S3. 122-1986.

Digital Representation for Communication of Illustration Data: CGM Application Profile, MIL-D-28003, 20 December 1988.

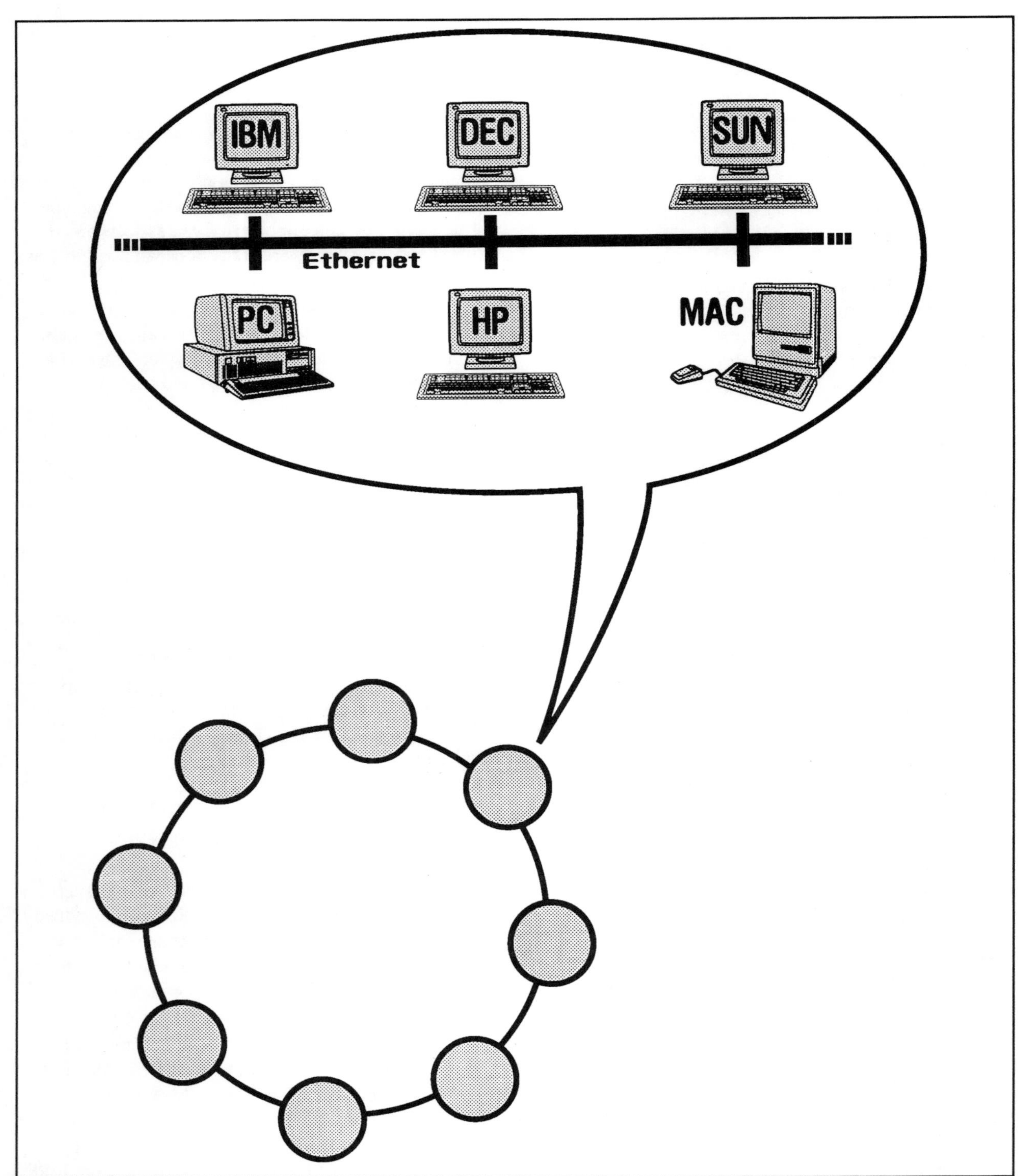

***Figure 1**. Eight network clusters serving 40,000 users.*

Presented at the Standards of CIM Conference, Febrary 21-23, 1989

An Overview of Electronic Data Interchange Standards

LESLIE ROHDE
The LJR Group

INTRODUCTION

There is no shortage of standards for the description of electronic products. This is not surprising, as standards are made by people from diverse backgrounds who have very different problems they want standards to solve, and who have various approaches to defining those standards.

The complexity inherent in the delivery of electronic systems, from concept to product, manifests itself in business structures and equipment that are both intricately structured and highly compartmentalized. In any manufacturing firm there are multiple, often competing, groups of people working to deliver the company's products. Within each group techniques and supporting tools have been developed that satisfy the unique viewpoint of that group, generally in isolation from the other groups in the company. As our industry matured, separate splinter cultures mirroring our business architecture formed associations and cooperatives to develop techniques and standards. The list of organizations which define, support, or require the use of standards for electronic products is extensive. The list below includes only some of them.

American National Standards Institute (ANSI)
Computer Aided Manufacturing-International (CAMI)
Department of Defense (DoD)
EDIF Volunteer Organization
Electronic Industries Association (EIA)
European CAD Integration Project (ECIP)
IGES/PDES Volunteer Organization
Institute of Electrical and Electronic Engineers (IEEE)
Institute for Interconnecting and Packaging Electronic Circuits (IPC)
International Standards Organization (ISO)
National Institute of Standards and Technology (NIST)
PDES, Inc.
U.S. Air Force

This paper will explore the four dominant standards in the electronic products arena: EDIF, IGES, IPC-D-35x and VHDL. (1a) EDIF, IGES, and IPC are each able to interchange a wide class of functional and physical product data. Thus, they have many overlapping features. The overlap of VHDL with the other standards is limited because it is intended primarily as a behavioral design language and makes no attempt to capture physical data.

IPC

The Institute for Interconnecting and Packaging Electronic Circuits (IPC) was originally conceived to support the printed circuits industry with development and dissemination of guidelines for manufacturing and assembly processes. The development of standards for the interchange of data in digital form is only a small part of what the institute does; however, historically it has been very important to its membership, who are generally from large electronic systems manufacturers, and this led the IPC to its early involvement in the standards area.

The ANSI/IPC-D-350 (1,2,3,4) series of standards is the oldest attempt at a standard format for PCB physical design data. The initial release of D-350 in 1972 supported board artwork data. The D-351 and D-352 standards released in 1985 added support for drawings and net-list data. The formats are recognized for use under DoD contracts, and this has been the primary motivation for implementation by CAE vendors. With the advent of other organizations whose sole charter is the development of data interchange standards, I expect the IPC will begin to take a participatory rather than a leading role. The user community has sufficient investment in IPC formats that commercially available bridges to other standards can certainly be expected.

IGES

The Initial Graphics Exchange Standard (5) was originally developed from the need to make the information from different CAD systems available in one neutral format. Membership is largely from government organizations, prime contractors, and aerospace firms.

The initial release of IGES in 1980 was only able to support mechanical data, but even at that time the electrical applications committee was addressing requirements for electrical product support. Following a false start in the second release, the third release finally provided an electrical entity subset which functioned as desired. The integrated description of electrical and full mechanical properties is a key aspect of the IGES format. The Figure 1-1 in the IGES Electrical Application Guide, (6) shows an example of the intended IGES application where a signal is described that passes through cables, connectors, printed wiring boards, and integrated circuits. This level of integration is not supported by any other existing standard.

While integration of mechanical and electrical properties is the primary advantage of IGES, there are few implementations of the electrical entities due to a misconception in the CAE vendor and user community that IGES is only for mechanical data. IGES is currently identified for use in DoD contracts and this has been the prime motivation for the few CAE implementations that do exist. It is unfortunate that IGES Electrical has not been better received by the CAE vendor community, because the electrical subset is very capable, and the IGES/EAC has done a good job of incorporating past experience in the product.

The Product Data Exchange Standard (PDES) is intended by the IGES organization to address long-term requirements—a limitation seen to exist by the organization with its current standard. It is interesting to note that the other standards discussed here are similarly limited in their ability to address future requirements. PDES is taking a formal approach wherein the conceptual requirements are separated from issues of format. Like IGES, PDES will eventually address the representation of all product data, mechanical and electrical alike; however, it has already provided a major contribution to the development of standards by its emphasis on the identification of the fundamental conceptual structures underlying the representation of a product in any format.

PDES has already been identified by the DoD as a key enabling technology for the CALS program. This government support has led the other standards organizations to adopt a more formal approach to facilitate integration with PDES. The conceptual structure developed by PDES for electronic products (18,19,20) can be applied in large measure to existing interchange formats, such as EDIF (28) and IGES.

EDIF

The Electronic Design Interchange Format (7,9) was born of a need for IC vendors to exchange IC mask data, but the organization has grown to include applications outside its original narrow purpose. The EDIF effort is the newest attempt at product data representation, so it stands to learn from the mistakes of the earlier efforts; however, it is also the least mature. While the IC manufacturers are still strong participants in the EDIF standards effort, the most active subcommittees are the PCB (8,11,12,14) and schematic subcommittees. A significant limitation of EDIF is the failing to recognize that electrical products are ultimately constructed of mechanical objects.

The PCB application guide (10) has been published, and experimental use of EDIF for PCB library and layout archive and transfer has been started. Several members of the PCB subcommittee are already constructing applications following the guide.

It can be expected that the EDIF standard will not officially incorporate the PCB application conventions until late this year. However, because of the extensive CAE vendor involvement, reliable interfaces will exist before the standard is updated. Interchange of schematics in EDIF among multiple competing vendor systems has been demonstrated. (1b) This is a significant achievement in light of the short time the standard has been in existence. In large measure, the success of EDIF is not a feature of the standard but of the commitment of the member companies.

VHDL

The VHSIC Hardware Design Language (15,16,17) (VHDL) originated from a need to supply circuit structure and behavioral data in support of the Very High Speed Integrated Circuit program, but it is being rapidly adopted by the entire logic simulation community. There is no provision and no intent to provide physical implementation data, so overlap with the other standards discussed in this paper is limited. VHDL is a design language and functions as an interchange format only as a byproduct of this goal; just as a schematic is initially a design tool, but may later function as product sustaining documentation.

Because hardware design languages are fairly recent, we will continue to see schematics used in functional design for some time. Hardware design languages will be found in more and more applications and can be expected to nearly replace schematics over the long term.

VHDL does overlap with other standards in the definition of functional structure. A design done in VHDL will include functional blocks that must be mapped to physical components implying an interface with EDIF, IGES, or IPC.

COMPARING STANDARDS

There are great differences in format between these standards, but format is not really an important consideration. A change in format is a problem if it requires new software investment, but the most important aspect of a transfer standard is what data and what applications it supports. With this goal in mind, the argument over which format is "better" is senseless.

The comparison of overlapping standards requires a meta-language (24) to describe the expressive power and application coverage of each standard at a conceptual level rather than at the level of syntax. Methods, such as IDEF1X, (21) VDM, (26) and the ECIP Conceptual Modeling Method (27) could be applied, but they have not been used to compare the existing standards. (1c) One of the barriers to the application of these techniques is the lack of conceptual models for the existing standards. Each of the standards uses an informal set of definitions, and the model, if one exists at all, is maintained in the syntax of the standard rather than in a neutral form accessible to independent review.

Within the EDIF PCB-TSC and the IGES/EAC we are attempting conceptual models, which should enable the formal comparison of these standards to be made in the near future. I know of no modeling project addressing the IPC standards.

So, with no formal foundation, the following ad hoc categorization must suffice.

Transfer of Functional Design

Currently, the functional design of electronic devices may be described by a schematic or a VHDL file. The use of schematics as a design vehicle is declining as more engineers adopt hardware design languages; however, this will be a long process. Whether a schematic or VHDL is used, the end result of the functional design process is a net list, which may be quite adequately expressed using any of the EDIF, IGES, or IPC standards. Commercially available bridges from VHDL to EDIF are described in the literature. (30,31,32)

Net list and schematic transfer in EDIF are common today and will continue to be the method of choice due to extensive vendor support. Few if any IGES net list and schematic transfer products exist, although the standard is as capable as EDIF and has been in use much longer. IPC also supports schematic and net list transfers, but I know of no implementations. Most IPC use is for PCB artwork, with the net list and drawing data having arrived too late to obtain a strong following.

Physical Design Interface to Manufacturing

The transfer of printed circuit board manufacturing and assembly data has been successfully supported by the IPC standards for some years. IGES and (recently) EDIF also support PCB manufacturing and assembly data. Each of these standards has strengths and weaknesses, but they are approximately equivalent.

ECAD to ECAD Transfer of Physical Design Data

The transfer of physical design data between dissimilar ECAD systems is the most difficult type of data interchange. The transfer must not only preserve product data, but must also provide for the transfer of data required by CAD systems. The only standard that is expected to succeed in this type of transfer is EDIF, for reasons having little to do with the standard—it is simply the one that ECAD vendors have chosen to adopt. (1d) IGES and IPC do not have the vendor support to make this possible.

Today, transfer of PCB physical data between ECAD systems is an ad hoc exercise. Owing to the relative youth of most ECAD products, features other than additional interfaces generally take priority, so progress on this has been slow. With the current instability in the ECAD marketplace, however, this issue may get more attention as customers come to recognize the strategic value of the data locked up in vendor formats.

HARMONIZING STANDARDS

IGES, IPC and EDIF share a significant area of overlap. Each of these standards is being created by separate groups of volunteers working in limited spare time. In each case the creation and enhancement of the standard requires two interwoven activities: conceptual modeling and language development. The language formats developed by these standards groups are quite different, but these differences are of little importance. The real expense in creating a standard is in the understanding of concepts and their interrelationships. The conceptual structures supported by the IGES, IPC and EDIF standards were generated in isolation from one another and yet are very much the same. The duplication of effort is immense! It is to everyone's advantage that these groups should work together on the conceptual issues and handle the language issues separately. (22,23) To a small extent this has been started between the IGES/PDES EAC and the EDIF PCB-TSC through crossover membership and correspondence, but more cooperation is needed. If consensus is achieved at the conceptual level, the integration of the standards, the development of PDES and the development of the Engineering Information System (29) would all benefit.

It is possible that this harmonization effort will very soon be mandated. A study of the state of electronic standards and how they are to be harmonized is under way by the CAMI/EAP under the auspices of ANSI. It is too early to know what form harmonization will take, but some of us are taking a "grassroots" approach in anticipation of some formal process.

GUIDE TO IMPLEMENTERS

In general, there are niches where the choice between EDIF, IGES, and IPC is clear, but there are also vast areas of overlap where the decision is less obvious. By contrast, the integration of VHDL into the design environment is less controversial because the overlap with the other standards is well defined. None of today's standards is a perfect choice, so much of the decision will depend on customer data requirements and formats already in use in your company. The use of standards yields several advantages:

- Protects your investment in data. (1e)
- Increases the availability of data for other processing.
- Reduces software investment through code reuse and an increase in data "shelf life."
- Promotes business integration through data availability and commonality of data formats.

But in all cases there are some general cautions that should be observed:

- Get involved in the standards process. Each of the standards discussed has an active user's group and the standards process itself is generally open.
- Do not put one of these standards in the critical path until it is proven within your company, for your intended application, and using a live pilot project.
- Plan for bridges to vendor formats and other standards.
- The big payoff comes from integration across business functions, not within a single function. But with big payoffs come big risks, so keep your expectations low initially.

Is it worthwhile? Yes. Useful work can be done with all of these standards today and the benefits of adopting a standard format of some kind can be substantial.

REFERENCES

1. ANSI/IPC-D-350B, Printed Board Description in Digital Form, IPC, 3451 Church Street, Evanston Il 60203, Aug 1977.

2. ANSI/IPC-D-351, Printed Board Drawings in Digital Form, IPC, 3451 Church Street, Evanston Il 60203, Sept 1985.

3. ANSI/IPC-D-352, Electrical Design Data Description for Printed Boards in Digital Form, IPC, 3451 Church Street, Evanston Il 60203, Sept 1985.

4. Dieter Bergman, "IPC-D-350: One Standard for PC Design," PC Design, Dec 1987, pp 17-18.

5. American National Standards Institute, Initial Graphics Exchange Specification (IGES). Version 3.0, ANSI, 1985.

6. Initial Graphics Exchange Specification Version 3.0 Electrical Application Guide, IGES/EAC, 1985.

7. Richard Olson, "Evolution of EDIF," Digest of Technical Papers, The Fourth EDIF User Group Workshop, Sept 1988.

8. Harvey Clawson, "Using EDIF for PCB Design," PC Design, Dec 1987, pp 19-32.

9. EDIF Steering Committee, EDIF Electronic Design Interchange Format Version 2 0 0, Electronic Industries Association, 2001 I Street NW, Washington D.C. 20006, May 1987.

10. Using EDIF for Printed Circuit Board Applications, EDIF PCB Technical Subcommittee, Draft Version, September 1988.

11. Harvey Clawson, "Using EDIF for Printed Circuit Board Applications," The Fourth EDIF User Group Workshop Digest of Technical Papers, EDIF User Group, 2222 South Dobson Road, Building 5, Mesa, AZ 85202, 1988.

12. Wayne Angevine, "Netlist and Schematic Transfer for PCB Applications," The Fourth EDIF User Group Workshop Digest of Technical Papers, EDIF User Group, 2222 South Dobson Road, Building 5, Mesa, AZ 85202, 1988.

13. Leslie Rohde, "Representation of PCB Library Data in EDIF," The Fourth EDIF User Group Workshop Digest of Technical Papers, EDIF User Group, 2222 South Dobson Road, Building 5, Mesa, AZ 85202, 1988.

14. Harvey Clawson, "Using EDIF for Physical Layout of Printed Circuit Boards," The Fourth EDIF User Group Workshop Digest of Technical Papers, EDIF User Group, 2222 South Dobson Road, Building 5, Mesa, AZ 85202, 1988.

15. IEEE Standard VHDL Language Reference Manual, IEEE Std 1076-1987, Institute of Electrical and Electronics Engineers, 1987.

16. John Hines, "Where VHDL Fits within the CAD Environment," PC Design, Dec 1987, pp 8-12.

17. Richard Goering, VHDL Tool Sets Support Design Verification," Computer Design, July 1988, pp 36-37.

18. Curtis H. Parks, Tutorial: Reading and Reviewing the Common Schema for Electrical Design and Analysis," Proceedings of the 24th ACM/IEEE Design Automation Conference, 1987, pp 479-483.

19. IEEE/PDES Cal Poly Task Team Final Report, April 30, 1987.

20. IEEE/PDES Cal Poly Task Team Extension Final Report, March 18, 1988.

21. Information Modeling Manual (IDEF1) - Extended (IDEF1X), ICAM Architecture Part II Volume V, Revised Dec 1985.

22. Ronald Waxman, "The Design Automation Standards Environment," Proceedings of the 24th ACM/IEEE Design Automation Conference, 1987, pp 559-561.

23. Larry O'Connell, "Design Automation Standards Need Integration," Proceedings of the 24th ACM/IEEE Design Automation Conference, 1987, p 562.

24. M. L. Brei, "Needed: A Meta-language for Evaluating the Expressiveness of EDIF, IGES, VHDL and other Representation Mechanisms," Proceedings of the 24th ACM/IEEE Design Automation Conference, 1987, p 565.

25. The UK EDIF Support Project (CAD/031), Final Report: EDIF Data Model, Part 4 - Data Model of EDIF, 1986.

26. David Chalmers, "Modeling EDIF V 2 0 0 Using VDM," The Fourth EDIF User Group Workshop Digest of Technical Papers, EDIF User Group, 2222 South Dobson Road, Building 5, Mesa, AZ 85202, 1988.

27. European CAD Integration Project (ECIP), The ECIP Conceptual Modeling (CM) Method Reference Manual, Reference ECIP.PH.D.032.02.

28. Leslie Rohde, "Compilation of EDIF Function Cells to the CPTT Model," Unpublished - available from the author, Oct 1988.

29. David Smith, "An Engineering Information System," VLSI Systems Design, Nov 1988, pp 60-68.

30. Eric H. Dashman, "A VHDL-to-EDIF Translator," The Fourth EDIF User Group Workshop Digest of Technical Papers, EDIF User Group, 2222 South Dobson Road, Building 5, Mesa, AZ 85202, 1988.

31. Alfred S. Gilman, "Using VHDL and EDIF in Concert," The Fourth EDIF User Group Workshop Digest of Technical Papers, EDIF User Group, 2222 South Dobson Road, Building 5, Mesa, AZ 85202, 1988.

32. Moe Shahdad, "An Interface Between VHDL and EDIF," Proceedings of the 24th ACM/IEEE Design Automation Conference, 1987, pp. 472-478.

33. C. Parks, P. Nelson, Logical Schema for Schematics for Electronic Products. IGES/PDES Electrical Applications Committee, October 9, 1985.

FOOTNOTES

1a. Listed alphabetically, but described by date of first release: one must be mindful of biases in this subject.

1b. The demonstration has occurred twice to date. First in the EIA booth at the 1988 Design Automation Conference and subsequently at The Fourth EDIF Workshop held in September 1988. The DAC booth contained about six vendor systems and the EDIF World Demonstration a dozen.

1c. Some models of existing standards do exist, but most are not conceptual or are incomplete. (13,25,26,33) The EIS program (29) has done some modeling of EDIF and VHDL, but only at the level of language syntax, not concepts.

1d. To be fair, users have been requesting EDIF support from vendors for the last two years. I'm not sure who started it, but there has been a feeding frenzy surrounding EDIF that is somewhat unjustified. On the other hand, the rate of progress on the standard has benefited greatly from the level of demand felt by CAD vendors.

1e. For most manufacturing firms, the investment in data exceeds the value of the physical plant.

Reprinted from the CIM Implementation Guide, 3rd Edition

PDES: The Enterprise Data Standard

ROBERT CARRINGER, CMfgE
Institute of Business Technology

INTRODUCTION

The Product Data Exchange Standard (PDES) is a data description and format standard under development for the exchange and sharing of data needed to describe a product and its manufacturing processes. This paper provides a general introduction to the technology, development plans, organizations, and benefits of PDES.

WHAT IS PDES?

PDES can be viewed from several different viewpoints and is actually a compilation of many different activities. PDES is a standards development process. It is a combination of different technologies, contributed by people from many companies and organizations.

As a standards development process, PDES is an activity with a goal of creating an international standard for the exchange of product model data. The resulting standard is also a process whereby knowledge is created, shared, and documented. The standards development process is unusual because the standards activity is happening concurrently with the development of the technology. In most standardization processes, the technology is developed, stabilized, and then standardized. The PDES program is doing these concurrently.

The technology going into the development of the PDES program can also be categorized in several different groups.

The first technology to be discussed is information modeling. Here, the purpose is to define and document the product data to be shared or exchanged between organizations and functions. Information modeling has become the modeling of knowledge, which includes:

- Knowledge about the enterprise.
- Knowledge about the products produced by the enterprise.
- Knowledge about the process used to produce those products.

Once the information has been modeled, there must be a way of storing and retrieving it. Information management technology is also an important part of the development of PDES. Information management is being stressed because of the different information used in the PDES program. Information management technology has to provide access to the PDES information, storage and retrieval of the PDES data, and the ability to create new PDES data and modify it.

The third category of technology is downstream technology that uses PDES data. PDES is frequently tied to downstream technology to demonstrate its capabilities. For example, a computer-aided process planning system, an example of downstream technology, is often an information starved application needing a PDES database to become fully automated. From the history of the development of PDES, we find that downstream applications provide excellent sources for the requirements of a PDES database, or a PDES system. Those downstream applications like computer-aided process planning, group technology, classification and coding, machine parts programming, robotics programming, inspection system programming—all require PDES data as input, and can also store back data into the PDES database.

These are all groups of technologies which together become part of the PDES program, and upon which the PDES program relies, and through which the PDES program will be demonstrated to the manufacturing and engineering communities.

PDES is also an activity. The PDES program started in 1984 as a spin-off of the IGES activity. It has been administered by the National Institute for Standards and Technology (NIST). NIST administers the PDES program through its IGES/PDES organization. The people involved with PDES are from companies throughout the world. They volunteer their time and company resources to work on the development of the PDES specification and on the standardization of PDES.

The activity of PDES can also be traced back to technology development projects funded by the Air Force Manufacturing Technology Program. Research and development programs have been funded by the Air Force since the early 1980s working on the technologies underlying PDES and other activities related to PDES.

When asked, "What is PDES?" we get several answers: PDES is a standards development process; a goal of the PDES program is to create an international standard for the exchange and sharing of product model data; PDES is also a technology; PDES is a technology of information management, modeling, user technologies, and implementation technologies related to the overall industry of CAD/CAM and CIM. And, finally, PDES is an activity contributed to by companies around the world, by U.S. government agencies, the Air Force, and other military branches of the Department of Defense.

These different viewpoints make possible this description and explanation of the PDES program, so that your company may become involved in the standards development process, understand the technologies related to the PDES program, and become involved in PDES, the activity.

The requirements for PDES and all its related activities are:

- Define the product and process of manufactured systems.
- Support exchange and sharing of product information with a minimum of human interpretation.
- Interrelate a broad range of product information to support applications found throughout the product life cycle.
- Define a single, logical representation of product information and application views (content and format).

The product and process information can be grouped into categories. The information is modeled using graphical and computer language techniques. The resultant topical data models contain information related to: geometry, solids, tolerances, electrical functions, material, presentation, architecture, topology, form features, layered electrical products, finite element modeling, product structure, drafting, and ship structures.

Information modeling is based on a three-layered approach. The logical layer is a definition of all the information outlined above. It is a single data model or representation of the data. Also called a conceptual layer, the logical layer is the central model from which the other two layers are derived. The application layer is a subset of the logical layer defining the information for a particular application (design, process planning, inspection, etc.). The physical layer is the definition of the actual file or database containing the PDES data. The PDES physical file format is an example of the physical layer.

A HISTORICAL PERSPECTIVE ON PDES

PDES has as its objective to electronically define all the information needed to design, manufacture, and support a product. We find some of the early demonstrations of PDES technology relate to being able to define the product's geometry in a computer system, demonstrate the automatic creation of machining instructions, and machine the product that fits that initial product definition geometry. That, as an objective, can be traced back to several things in the past, going as far back as the development of numerical control in the 1950s.

The original objective of the NC development program was to automatically define a part so that control of part manufacturing moved off the shop floor to an engineer in an office. The engineer could control how the machine ran and the products the machine made. That objective was partially realized in the development of APT. We find in APT the ability to define simple geometry and tool paths to manufacture specified geometry.

Defining products in APT was really limited to the simple geometry modeler within an APT program. It was not sufficient to support other downstream applications and other upstream applications to create the geometry in the first place.

Through the 1960s, we have seen the implementation of graphics systems within manufacturing organizations. These systems define and model products, and have become computer-aided drafting systems. There is also storage of information about a product. Throughout the 1960s and 1970s, the development of CAD/CAM evolved in such a manner. We are defining products, we are moving from the ability to put out an automated blueprint to being able to put out geometry information that can drive downstream applications like NC part programming. Once we obtain NC part programming, we want to use it for things like coordinate measuring machine instruction, and also upstream into the design process to be able to perform design analysis on this CAD/CAM data as well.

In the late 1970s addressing the need of exchanging CAD/CAM data between CAD/CAM systems, IGES was invented by Boeing, General Electric, and the Air Force. The objective of IGES is, again, the ability to exchange a product definition between two different systems. And why do we want to do that? Because we want this information about the total life cycle of the product to become digital and we want to exchange digital information or electronic product description data instead of paper drawings.

From our experience with IGES in the early 1980s, the manufacturing engineering community saw there was a need for other information to be stored electronically. Information like the manufacturing intent of a product, or the manufacturing form features, as we later called them. In viewing information as it is released from engineering to manufacturing, we see there are several categories of information actually released from design or desired to be released from design, to support applications and manufacturing.

The Air Force funded a program, the Product Definition Data Interface, whose objective was to define information needed by manufacturing, created in engineering, to support manufacturing applications.

The PDDI program presented its early results to the IGES organization in 1984. The IGES organization formed the PDES initiation activity. That was the first time the IGES organization chartered any activity with relation to PDES and, in fact, was the early genesis of the PDES program.

Design automation, manufacturing automation, and the implementation of computers to design and the manufacturing process have all created awareness of the need to share common information about a product and the process used to make it.

We can go back as far back as the 1950s to find activities which today can be related to the PDES program. The objectives, though changed with viewpoints of technologies, are very similar. The objective of automating the design and manufacturing process is to achieve the highest efficiency and create the highest quality products at the lowest possible prices.

That, again, is our goal with the PDES program.

The PDES activity is addressing the following items: the need for exchanging data in a global manufacturing environment; exchanging data between domestic and international trading partners; and exchanging data between manufacturing organizations within a corporation, between a corporation and its suppliers, and between a corporation and its customer.

While related technologies and the technology foundation have changed since the 1950s, we have very much the same objectives as we had then. With new technology we are able to address them again to a more complete stage. We find the PDES program to be addressing the identification, categorization, storage, exchange, and sharing of product data throughout the entire product life cycle from its early conceptual design through release to manufacturing, through delivery to the customer, and on through product support.

ARCHITECTURE FOR IMPLEMENTATION

PDES will be implemented to satisfy two primary needs: exchanging product data between different applications, and sharing product data between several different applications.

The PDES organization has defined four implementation architectures, two for exchange, and two for product data sharing.

Level I passive file exchange represents an implementation architecture where product data is transferred from one application to another in a batch process as seen in Figure 1. Level I is similar to the way in which IGES is implemented today. Application A creates product data in a file format as prescribed by the PDES organization. That file is then transferred to Application B. Application B has a post processor translator to post process that file into its own native data structure. This transfer can be bi-directional if both applications have a pre- and post-processor. Level I is focused on the exchange of product data between two applications. Level II has the same focus.

Level II active file exchange is represented by an application that creates product data in a neutral format. This process is outlined in Figure 2. The neutral format has a working form which is a memory resident model of the product data. The two applications, A and B, can reach into that memory resident model or working form model, and pull out entity by entity the product data it needs. In the exchange process for

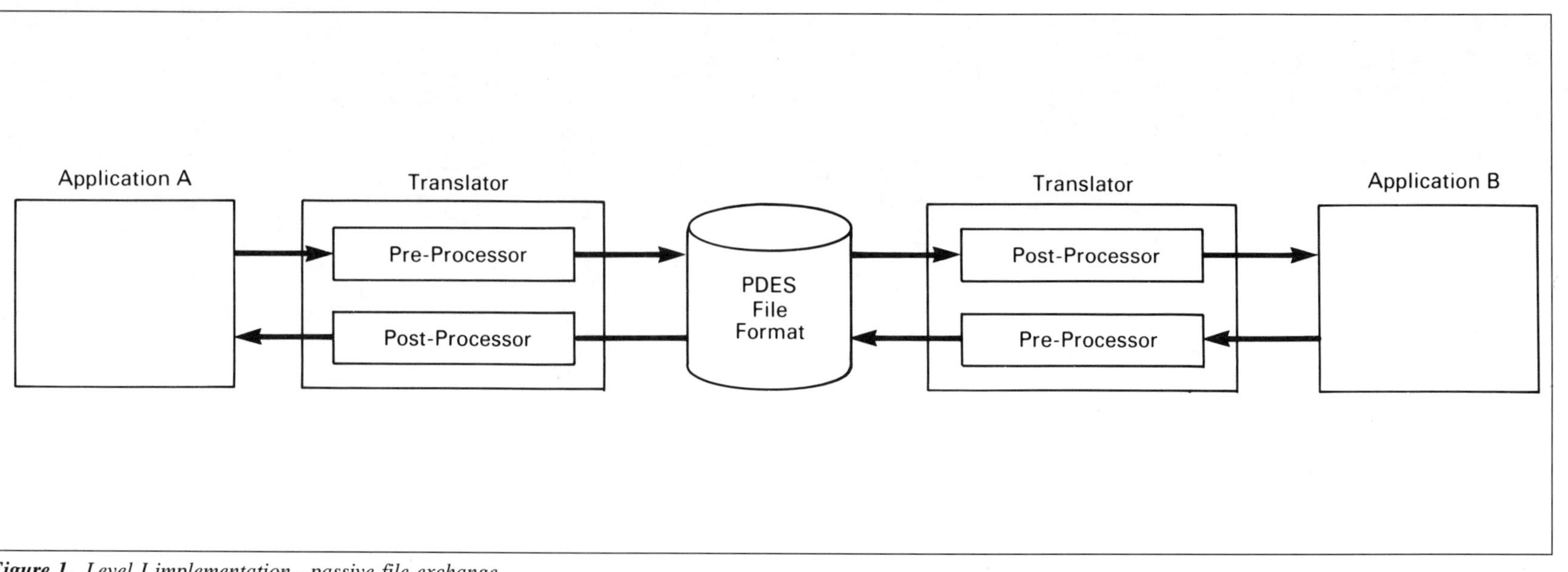

Figure 1. Level I implementation—passive file exchange.

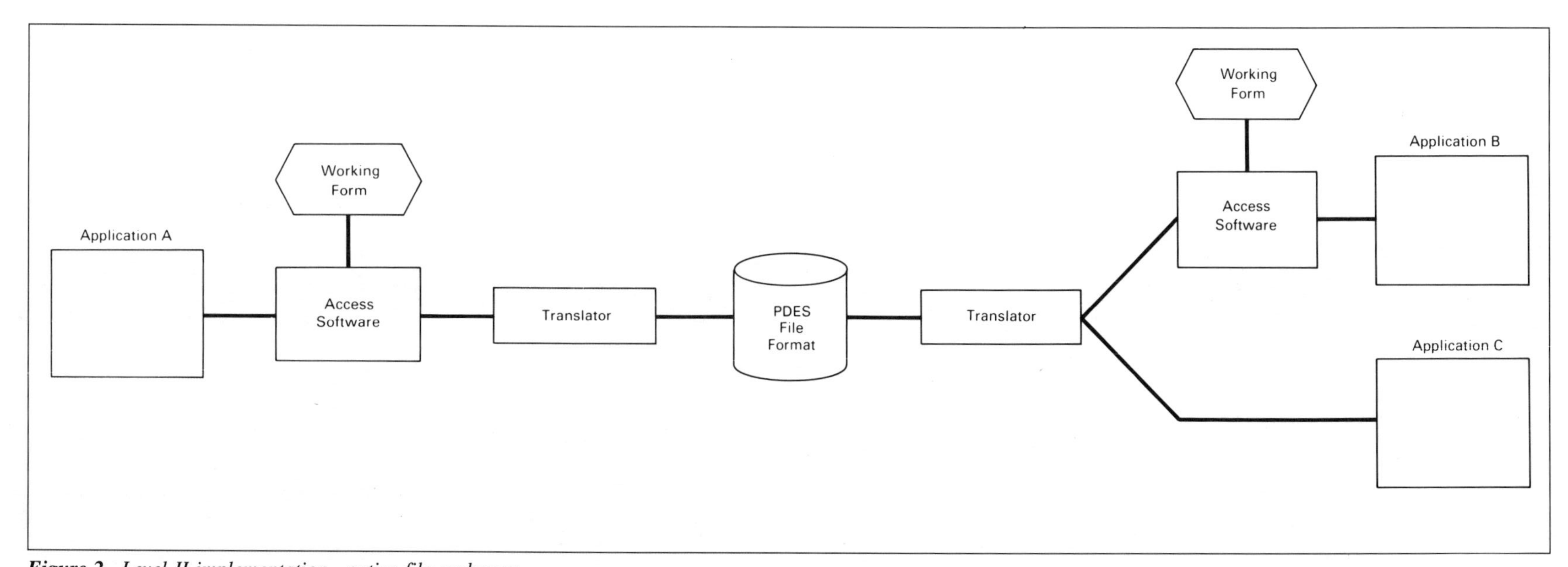

Figure 2. Level II implementation—active file exchange.

a Level II implementation, there are two representations of the product data: the exchange file format, or physical format, and another physical form called the working form, which is a memory resident model that can be accessed entity by entity instead of in the entire file.

An example of the Level II implementation is the Air Force PDDI (Product Definition Data Interface) program or GMAP (Geometric Modeling Applications Interface) program. Both programs used a neutral form in a physical file format, and a neutral form in a working form memory resident model format. Both Level I and Level II implementation architectures are focused on the exchange of product data between applications. Level III and IV are focused on the sharing of product data as seen in Figures 3 and 4.

Level III shared database implementation has a logical description of all the product data and a physical implementation of this information in a database format. Applications can then reach into that database and pull out the information they need to drive their specific application. An example would be a relational database with several applications driving from it. Using the data dictionary, which describes data in the database, the application then has the capability to ask the database manager for pieces of product data it needs to do its job and can provide back end resultant data to be stored into the product definition database. The Air Force integrated design system, or IDS system, developed by Rockwell International, offers a good example of this type of technology.

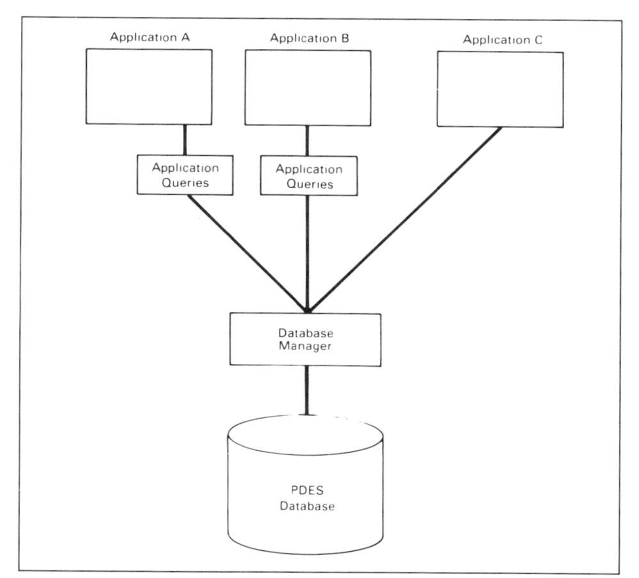

Figure 3. Level III implementation—shared database.

A Level IV knowledge-base implementation of PDES is represented by using an object-oriented approach to the storage of product definition data where the product data and methods to operate on that product data are stored together as an object in an object-oriented database. This knowledge-base representation is the desired implementation environment, and is a limited prototype implementation of a Level IV as it is performed today. Currently, Level IV is not widely implemented throughout industry and the vendor community.

PDES INC. PROGRAM

The PDES Inc. program is an industry-funded cooperative project to develop the PDES specification and supporting software. The objective of PDES Inc. is to accelerate the development and implementation of PDES within industry. The program is using a combination of industry-provided personnel knowledgeable in PDES technologies, and hired contractors working together focusing on the development of the PDES specification.

The background of the PDES program dates back to the fall of 1986. Individuals from a few aerospace companies worked with the United States Air Force Manufacturing Technology Division at Wright Patterson Air Force Base to develop plans for a program to accelerate the development of PDES. Looking at several alternative methods for funding the overall development program, these companies decided to focus on an industry-only membership, where government participation would be in a nonfunded involvement.

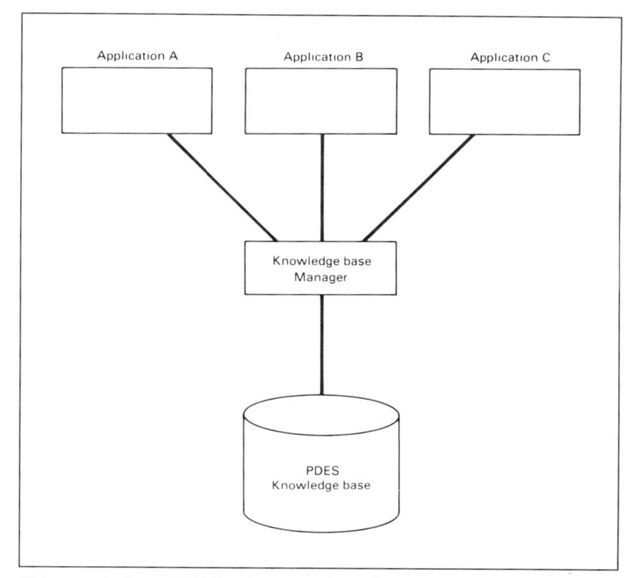

Figure 4. Level IV implementation–knowledge-base.

This activity grew until the development of a interim PDES Inc. planning activity, formulating the initial technical development plan for the PDES Inc. program, and managing the release of a request for proposal for a PDES Inc. host contractor.

The South Carolina Research Authority teamed with Battelle Memorial Laboratories, Dan Appleton Company, International Technegroup Inc., and Arthur D. Little to create a proposal to be the host contractor and to provide technical resources for the management of the technical work to be done on the PDES Inc. program. The PDES Inc. board of directors awarded the contract to the South Carolina Research Authority and its technical team in August, 1988. The contract officially was kicked off in October of that year, with a team meeting of the initial professionals from each of the member companies and technical subcontractors involved in the project.

The PDES Inc. program is divided into two major phases, each lasting 18 months. In Phase one, the objective is to develop a Level I or Level II implementation of PDES. In Phase two, the objective is to develop a Level III implementation of PDES.

The team of people divided itself into three working groups: the model integration team, the model test and validation team, and the technical products/implementation team. This is illustrated in Figure 5. In each group, a team leader was chosen from one of the member companies. A technical subcontractor and personnel from the different member companies were also assigned to each group.

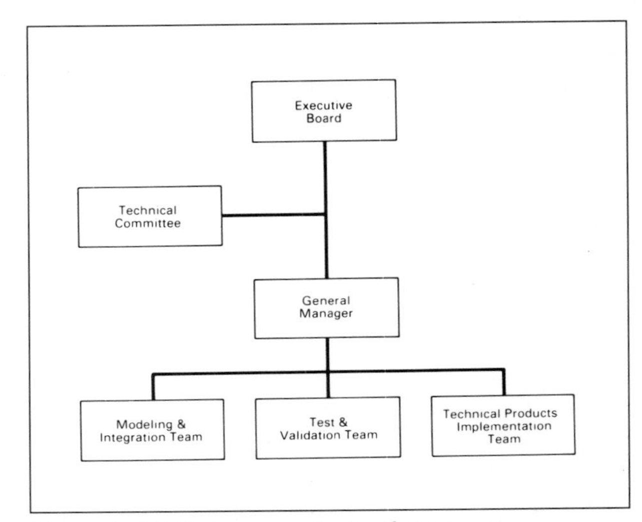

Figure 5. PDES, Inc. organizational structure.

An important philosophy expressed by the PDES Inc. program defines the relationship of its proprietary results with the PDES volunteer organization. In the spirit of cooperation, PDES Inc. will provide suggested changes to information models received from the volunteer group. The changes will be generated by the PDES Inc. process of modeling, integration, test, validation, and implementation of PDES data models. The process and relationship are depicted in Figure 6.

In addition to the three working groups, there was a configuration control board, and a systems integration board comprised of the host contractor general manager, the team leaders from the working groups, and the technical subcontractors. An important link was established between the PDES Inc. program and NIST National PDES testbed. NIST became the PDES test laboratory and provided personnel to the PDES Inc. program to help with testing and implementation. NIST received a contract from the Department of Defense to provide these types of services by establishing the PDES National Test Laboratory.

A company can choose to become a member at one of three levels: A Class I member, which is the high-

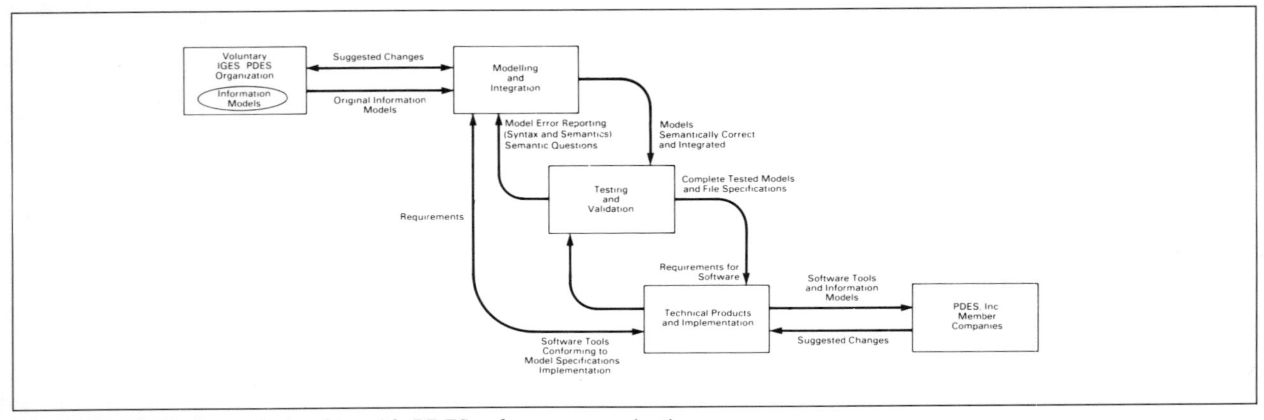

Figure 6. PDES, Inc. relationship with PDES voluntary organization.

est class, pays $100,000 per year plus provides two technical people for the working groups. In addition there is a $50,000 travel and equipment budget provided to each of the technical people. Class I members have membership on the board of directors as voting members, and own all rights to all deliverables in the PDES Inc. program.

A Class II member pays $50,000 a year, provides one technical person to the working groups, has a seat on the board of directors in a nonvoting capacity, and has the same rights of ownership to the results of the PDES Inc. program.

The third class is an observer class. The fee is $25,000 a year. There is no requirement to contribute technical labor, but there is also no ownership in the results of the PDES Inc. program.

The following table shows the membership as of September, 1989.

The PDES Inc. program will deliver results in a time frame which can be incorporated into the CALS Phase II effort. The Department of Defense, although not funding the PDES Inc. program outside of funding NIST, is supporting the PDES Inc. program by providing projects to companies that are involved in PDES Inc. The projects require the use of PDES in the delivery of data to the government. They are also sending encouragement to companies who have not yet joined the PDES Inc. program.

The planning people in charge of the CALS activity at the Department of Defense see the PDES Inc. program as providing PDES results to be used as the cornerstone of the data content and data delivery in a CALS Phase II environment for the future.

THE BENEFITS OF PDES

Looking at manufacturing from a bottom-up functional view, we find the introduction of computer technology can automate several of those discrete functions. Each implementation of automation itself becomes a small island of automation. An automated function typically has involved with it a user interface, some kind of functionality or algorithm coded into a software program, data input, data output, and data storage. The data repository that becomes the culmination of running this automated function several times becomes a small database.

The spectrum of manufacturing functionality from design through manufacturing to product support, has a great deal of information about our business stored in discrete implementations of technology. Implementations that are not interchangeable and cannot be integrated, can reside on top of other databases, or use information generated by other automated functions.

This is the problem we face today as we attempt to implement new technologies that may change a particular functional application. Every time we do this, we face investment in changing our technology. This is expensive and time consuming.

As an industry, we need to store information in a form we can retrieve year after year and still use, even if it is in a different form of automated functions. With the rate of change we have today in technology, we are unable to accomplish this. We cannot, for example, come back in five years with the same CAD/CAM system and expect to retrieve the same engineering model that we had. The technology changes too fast. With new technology, we eliminate paper. We have paper engineering drawings that are 100 years old. Paper has been an outstanding form of

Table 1.
PDES Inc. program membership.

Class I	Class II	Class III
Boeing	LTV Aerospace	Honeywell
General Dynamics	Rockwell	
General Electric	Prime Computers	
Grumman	DEC	
Lockheed	FMC	
McDonnell Douglas	Westinghouse	
Northrop	Newport News Shipbuilding	
IBM		
Martin Marietta		
General Motors		
United Technologies		

interface between various functional applications. This is because a human was involved with each functional application.

Now, we have automated applications, computer-aided design/computer-aided manufacturing, engineering analysis–all of these are done on computer. Inspection, too, is done with computer-driven machine tools. Automated assembly and automated material handling have as a function some computer automation, which requires information input and generally creates information output.

The total sum of this information is what we have defined earlier as product data: data that defines the product and the process used to manufacture it.

We've made attempts before to define neutral formats for storage and retrieval. One significant attempt is IGES. And, although IGES is used very heavily in the exchange of engineering information between two different CAD systems, it has not been successful in long-term storage and retrieval. IGES simply was not designed to do that. It contains a file structure and a format which does not provide easy storage. The PDES organization needs to be considering that. PDES needs to be addressing the implementation levels that we have discussed between exchange and sharing of information. It also needs to be addressing the long-term storage and retrieval of product data.

The importance of PDES from a business viewpoint is that it will provide not only a format for storing, archiving, and exchanging information, but the definition of that information. This will allow us to create automated applications which link directly to the PDES structure, and which can use PDES information to perform their functional capability.

It is important for those of us involved in the development and use of PDES to recognize the limitation and scope of PDES. PDES as a technology will provide the capability to exchange and share information. PDES as a technology does not answer all the questions faced by information management. While PDES addresses the information flow into and out of all the functional applications within a company, it does not address the organization of the company's handling of information, nor does it address the management of information within the company.

It is up to the manufacturing professional to address the capability of handling information within individual companies. The format and content of that information can now be changed because of the introduction of PDES. With PDES, we can handle the information in an electronic form, define enough structure into it to fill it with as much intelligent product and process information as possible, and still drive all the different functional applications in the manufacturing enterprise.

The information management within your own business may yet become another functional application within your manufacturing enterprise. As you approach what has been called a computer integrated enterprise, capabilities provided by a PDES specification and PDES implementation software will assist you in implementing new technology to achieve an overall integrated enterprise.

Presented at the CASA/SME AUTOFACT '88 Conference, October 30 - November 2, 1988.

Computer-Aided Acquisition and Logistics Support (CALS) Primer

PAUL N. PECHERSKY,
E-Systems Incorporated

WHAT IS CALS?

CALS is a U.S. DoD initiative undertaken to achieve significant weapon systems life cycle cost savings. These savings will be attained by applying existing and emerging computer and communications technologies to improve productivity, quality, and timeliness in development and support of weapon systems. Basically, CALS is a strategy to evolve from the current paper-intensive acquisition and support processes to a highly automated mode of operation. In the early 1990s, CALS requirements will be imposed on weapon systems entering production. Once implemented, defense contractors will be required to provide deliverables, such as tech manuals and engineering drawings, in digital form, rather than paper. Ultimately, DoD would like on-line access to customer-owned computers which contain program-related technical information.

CALS is an initiative or strategy. It is not a system which can be bought off the shelf; not a piece of software; not a piece of hardware. It is oriented toward technical information only.

WHY CALS?

As DoD budgets have grown over the last several years, ways to moderate this growth while not impacting their overall mission are being considered. OSD personnel indicate they expect current budget constraints to increase the focus on CALS.

DoD estimates of CALS benefits are providing significant ammunition for achieving this goal. For instance:

- Tech manual authoring automation is expected to provide a 20-30% increase in productivity.

- On-line access to maintenance information is expected to yield a 35% improvement in trouble-shooting accuracy.

- The Air Force estimated a $135 million annual savings in tech manual changes.

- An integrated database is expected to reduce ILS costs by 20-35% and reduce overall acquisition costs by 5-10%.

- The biggest benefit estimated by DoD is a 20% savings in life-cycle weapon systems costs.

HOW DID CALS BEGIN?

In 1984, a joint DoD/industry task force was formed to study the feasibility of extending computer integrated manufacturing (CIM) productivity gains to the product support area. Results of this study indicated the primary target for automation and integration was the large volume of technical information required to support a weapon system throughout its life cycle.

The task force also found many examples in DoD and industry of technology to automate the preparation and storage of technical information in digital form. The task force concluded that the greatest payoff would be achieved if DoD and industry could integrate these islands of automation.

These recommendations became the basis for CALS. In September 1985, Deputy Secretary of Defense Taft issued a formal memorandum which officially started CALS.

The objectives of this memo were:

- To accelerate the integration of reliability and maintainability tools into contractor computer aided engineering and design systems.

- To encourage contractors to automate processes for generating logistic technical information.

- To increase, through the modernization of DoD systems, the capability to receive, distribute, and use technical information in digital form.

On August 5, 1988, Taft issued a second memo stating that "effective immediately, plans for new weapon systems and related major equipment items should include the use of the CALS standards." This directive also stated that: (1) program managers should review opportunities to apply CALS standards for systems now in full-scale development or production; and (2) for systems entering development after September 1988, the cost and quality implications resulting from the CALS strategy will be given significant weight in source selection.

WHAT IS THE SCOPE OF CALS?

The target systems for CALS are those with primary purposes of creating, modifying, storing, distributing or using weapon systems technical information.

These systems include:

- Technical database definition and access.

- Digital interchange of data for:
    - Product definition
    - LSA/LSAR
    - Technical manuals
    - Training materials
    - Technical plans and reports
    - Operational feedback

- Integration of processes such as R&M.

WHY DIGITAL INFORMATION DELIVERY?

Although CALS will focus on many issues, the most immediate and visible impact on contractors will be digital delivery of weapons system program deliverables. Among the reasons for this impact are:

- Technical manuals and training materials become out of date quickly and, in many cases, are difficult to use and maintain.

- Engineering drawings are incomplete, illegible, and impossible to control.

- Configuration management information is not documented well and is difficult to maintain.

- Logistic support data is voluminous, redundant, and excessive, and not readily available to users.

- Reprocurement technical data packages for spare parts replenishment are inaccurate and incomplete, and take too long to prepare.

How Will Digital Delivery Work?

Figure 1 shows digital delivery aspects of CALS. These aspects can be characterized as a "System of Systems," composed of three key elements:

- Contractor's systems, such as design, manufacturing, and customer support.

- Government systems, such as acquisition and logistics support.

- Interfaces between industry and government.

Information passes between these systems in the form of documents, processable files, and interactive access to databases.

The boxes at the bottom of Figure 1 delineate the two types of CALS standards required to obtain and efficiently utilize digital data in weapon system contracts: namely, interchange standards and functional integration requirements.

Interchange standards include: (1) File formats; (2) data; (3) media (tape and optical disk); and (4) manner of transfer (on-line, mail or electronic). CALS will focus on two different information exchange methods—bulk transfer and on-line access. Currently, bulk transfer is recommended for all final contractual deliverables. Magnetic tape is the easiest, lowest cost, and most standardized exchange method available today.

Functional integration requirements consist of contractual tasks used in statements of work which define the contractors' capabilities for integration of data systems and processes. For the performance of DoD contracts these requirements will specify the integration of design, manufacture, and support processes. A well-established principle is that up-front design attention offers the greatest leverage in controlling life-cycle costs. Consequently, issues of reliability, maintainability, parts interchangeability, and other attributes fall into the category of functional integration.

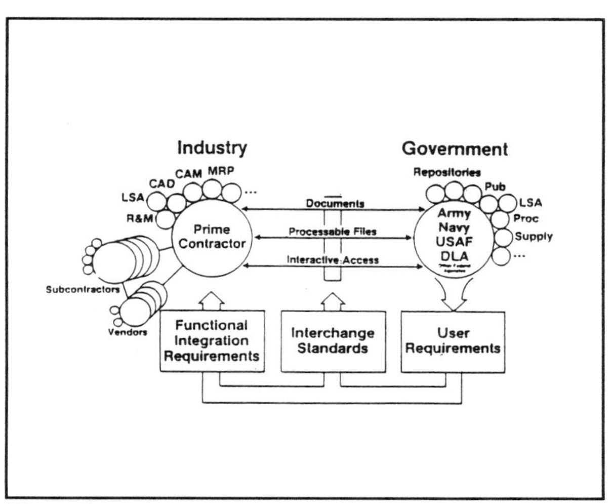

Figure 1.

WHAT IS DOD'S STRATEGY REGARDING CALS?

Early research indicated that, with or without CALS, advances in computer and communications technology would result in an environment where computers could exchange program information. The key was to insure that the required technology and systems were implemented in a planned and organized way with particular attention to integration. In other words, CALS is the mechanism to facilitate integration.

To avoid the "island of automation" syndrome which would have most certainly resulted without CALS, DoD developed the following core strategy:

- Facilitate and incentivize industry investment in integration of technical data systems and processes.
- Develop, test, and implement a phased series of standards for digital data interchange and integrated database access.
- Include CALS standards and integration requirements in DoD infrastructure modernization programs.
- Sponsor R&D in advanced technology for integration of technical data.

What is the CALS Acquisition Strategy?

CALS requirements will be phased in. Systems will be modernized. Demonstrations and prototypes will be developed, lead weapon systems have been chosen, and beginning in 1990, CALS requirements will become routine in contracts.

To minimize the method by which contractors electronically communicate with DoD, a core set of military interface standards is being developed. They will be used by all services including the Army, Navy, Air Force, and DLA. Figure 2 is a decision matrix for use by DoD acquisition managers to determine what CALS capabilities to specify in contracts.

The acquisition manager will be faced with four major decisions: (1) the type of deliverable—document, data file, or access to contractor-maintained computer; (2) the form in which data is to be delivered—hard copy, text file, graphics file, etc.; (3) the military standard to be specified for exchanging the data; and (4) should it be transmitted over telecommunications lines or delivered on some physical media, such as a magnetic tape.

How Will CALS Be Implemented?

The CALS policy office adopted a phased implementation strategy to take advantage of near-term capabilities to digitize paper and in parallel, develop advanced technologies and standards needed for a fully integrated, distributed information environment in the 1990s.

Initially, the focus will be on automation using available technology and will operate primarily in a batch environment. Digital information to be transferred will be put on magnetic tape using available standards and sent to DoD either electronically or using nonelectronic methods, such as express mail.

Phase II will focus on integration using advanced technology and standards. The central element in this phase will be an advanced product data model (PDES) and information will be accessed on-line from contractor or DoD computers.

Because computer aided design (CAD), automated authoring and publishing systems, and automated LSA databases already exist in many defense companies, Phase I will concentrate on a few high-volume technical data categories, starting with engineering drawings, technical manuals, and logistic support analysis data. Many of these systems are already capable of preparing data in digital form, and automated repositories are coming on-line in DoD.

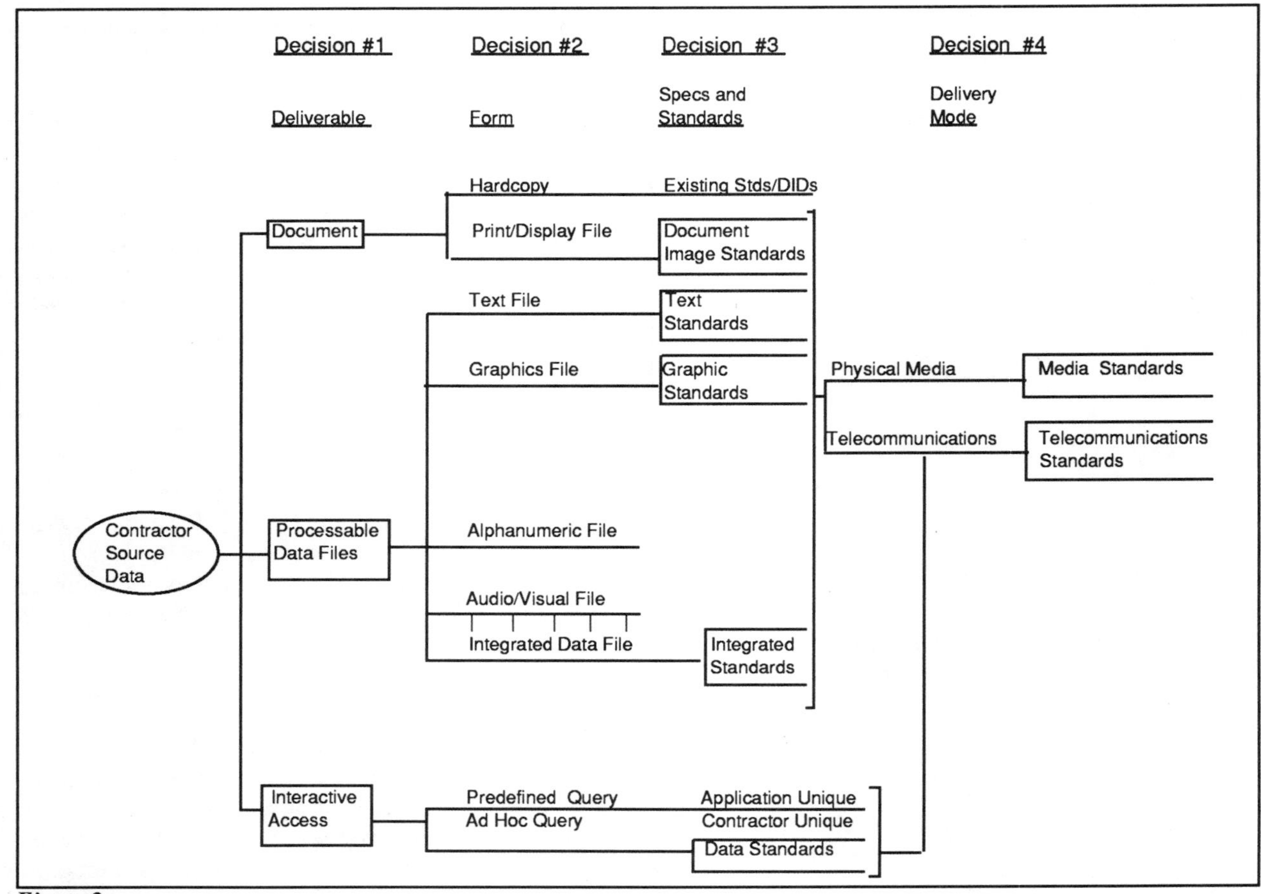

Figure 2.

Phase II is directed at integrated systems. This phase envisions utilizing new, emerging technologies and standards to allow different computers and applications to share common databases. This phase relies heavily on industry for the development of the required technology and standards.

DoD's strategy is to foster adoption of an industry-developed approach, rather than vice versa. DoD has already indicated strong support for the adoption of an industry cooperative to develop the product definition exchange standard (PDES).

When Will DoD Publish CALS Standards?

Figure 3 is DoD's plan for the release of CALS standards and the new functionality they expect to achieve. On December 22, 1987, DoD issued MIL-STD-1840A. Its purpose is to standardize the digital interface between organizations or systems exchanging digital forms of technical information. On the same date, MIL-D-28000 was issued to identify the requirements when product definition data is delivered in digital form using IGES (initial graphics exchange specification). This standard will initially focus on three product classes: technical illustrations, engineering drawings, and electrical/electronic applications. On February 26, 1988, DoD issued MIL-M-28001. This specification established the requirements for the digital data form of technical publications.

WHAT IS INDUSTRY'S ROLE?

DoD recognized at the inception that CALS would only work if it was a joint DoD/industry effort. As a consequence, the industry steering group depicted in Figure 4 was established.

More than 350 industry volunteers have organized into six working groups and numerous subgroups to develop technical inputs to CALS standards, coordinate CALS documents, and provide recommendations on management and implementation issues. In DoD's 1987 CALS report to congress, they stated that the

Road Map for Incremental CALS Core Releases

	1987	1988	1989	1990
Product Definition Data	Engineering Drawings • Raster • IGES (2 Subsets)	Engineering Drawings (Additional IGES Subsets)	Standards for Electronics Full Digital Data Package Contractor Maintained Data Base	PDES
Integrated Support Data Base	On-Line Access to LSA (Contractor Specific)	SOW Language to Integrate R&M with LSA	Initial Data Dictionary and Query Language Std. Functional Std. for IWSDB to Support LSA, R&M, Provisioning, TM Authoring, ISD, etc.	Expand Data Dictionary to be Compatible with PDES
Technical Manuals	Digital Capture for Auto. Publishing • SGML Tags • 1 TM Format • Raster Graphics • IGES (1 Subset)	Digital Capture • Additional TM Formats • CGM Graphics • Initial Paperless Delivery Format	Paperless TM's • Functional Std. for Authoring • Interface Std's (SGML, CGM, Raster)	Integrated Maint. Aid • Interfaces with Weapon System • Functional STD for "Intelligent" Maint. Assistance

Figure 3.

industry task force was providing an excellent vehicle for building the DoD/industry interface. The steering group, operating at the policy and executive level, is composed of the following working groups:

- Design integration, which focuses on reliability and maintainability, logistics support analysis, concurrent engineering, and design data capture. To date, this group completed a study regarding the implications of design integration within the electronics environment and is currently conducting a similar study for the nonelectronics arena.

- Security for both data and communications. Although they are looking at many issues, the major issue seems to be that of multilevel security and cost.

- Digital information interchange concentrating on the method and type of information to be transferred between DoD, prime contractors, subcontractors, and suppliers.

- Program acquisition will identify and offer recommendations regarding contractor incentives, contracts, legal issues, and benefits.

- Education and public communication will assist DoD in educating the contractor community about CALS.

- The international working group will focus on getting the non-U.S. based contractor community on the CALS band wagon. All DoD contractors and suppliers, whether foreign or domestic, must be CALS compliant.

Finally, in order to insure that the groups' activities are directed toward the right areas, a close working relationship with the DoD CALS steering group which represents OSD, the services, and other agencies is maintained.

Is Industry Prepared for CALS?

In June, 1987, at the request of OSD, the CALS industry steering group conducted a survey to establish a baseline of computer-aided technology in use by industry. The survey was sent to 40 of the top defense contractors, by dollar volume, for 1986. Thirty contractors responded.

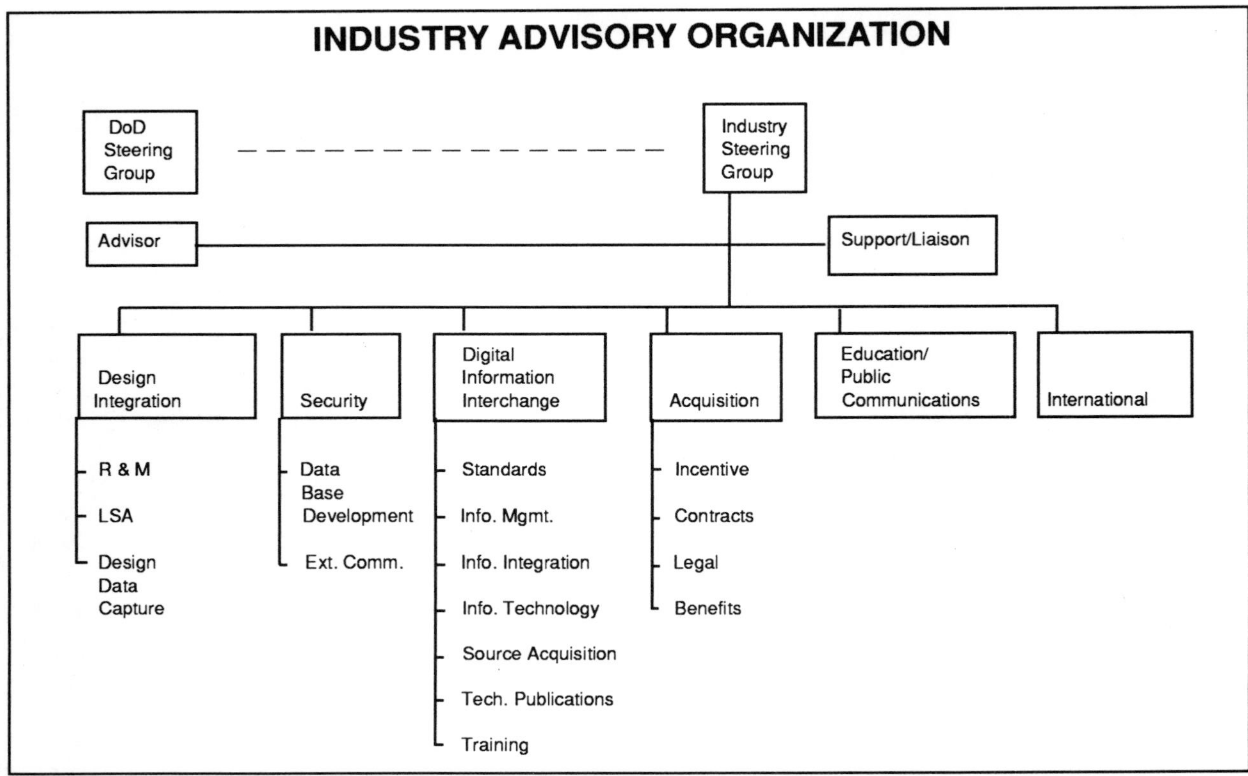

Figure 4.

The sample population was biased toward larger contractors: 78% had defense revenues in excess of $300 million, and 40% had greater than 10,000 employees. The 40 contractors surveyed accounted for 54.3% of the total dollar value of prime contractor awards in 1986.

The results were:

- All have implemented CAD, CAM, CAE and CIM technologies in various forms. The consensus is that it will take 10 to 15 years to achieve an 80 to 100% implementation of these technologies.

- Almost all already deliver some data in digital form to DoD customers.

- 40% have at least one third of their design, development, and manufacturing databases in digital form.

- 24% have integrated more than one third of their mechanical analysis data into mechanical CAD systems, while 34% have integrated more than one third of their electrical analysis data into electrical CAD systems.

- Although all use CAD systems, the majority of designs are still created manually. Only 18% created more than half of their drawings on CAD systems.

- In the electronics world, 83% use multivendor CAD systems. Seventy two percent use IGES for the transfer of electrical data. EDIF, VHDL (for VHSIC), and IPC are used by less than 25%.

- Only 21% are transferring more than 30% of their digital engineering data directly to manufacturing.

- In the mechanical CAD environment, 2D modeling is used by more than half, while 3D, surface, and solids are used on a more limited basis.

- None has achieved complete digital integration of R&M into the design process; however, almost half claim partial integration.

- Two thirds use CAD-created graphics to compose technical publications. While 80% have the ability to deliver technical publications in digital form, only 5% delivered any appreciable amount in digital form in 1986.

Will All Contractors Be Impacted by CALS?

DoD intends to make CALS routine in the way services acquire and support their weapon systems. They will incorporate CALS functional requirements and technical standards into weapon system contracts and into contracts for data systems developments that are part of the logistic information system infrastructure modernization effort.

With the completion of the Phase 1.0 specifications and establishment of additional repository capabilities, the services will begin selectively increasing the amount of digital delivered data for weapon systems that were developed prior to the CALS initiatives.

Digital data delivery for newly-developing weapon systems will commence with those whose full-scale development (FSD) phase falls in 1992-1995, based on contracts written in 1988-1989.

Finally, as the benefits of CALS become apparent over the next few years, competitive pressures will become a dominant factor. According to the June 30, 1987, report to Congress issued by the Office of the Assistant Secretary of Defense (Production and Logistics):

"Companies without CALS capabilities will not win contracts."

For further information regarding CALS, you may contact any one of the following individuals:

CALS DoD Steering Group

Dr. Michael McGrath,
Director, CALS Policy Office
Office of the Secretary of Defense (OSD)
202-697-0051

Mr. Bruce Lepisto,
Deputy Director, CALS Policy Office
Office of the Secretary of Defense (OSD)
202-697-0051

CALS Industry Steering Group (ISG)

H.B. Stormfeltz
Vice Chairman, ISG
Northrop
213-416-3106

Howard Chambers
Co-Chair, ISG Security Group
Rockwell International
213-414-1841

H.J. Correale
Co-Chair, ISG
Acquisition Group
McDonnell Douglas
314-234-7040

John Goclowski
Co-Chair, ISG Education and
Public Communication Group
Dynamics Research Co.
508-475-9090

William Jascomb
Co-Chair, ISG Security Group
Lockheed Georgia Co.
404-494-2625

Naomi J. McAffe
Co-Chair, ISG Design Integration Group
Westinghouse
301-765-3400

Joe Meredith
Co-Chair, ISG Design Integration
Group
Newport News Shipbuilding
703-892-4470

Stanley Meyers
Co-Chair, ISG Digital
Information
Interchange Group
Grumman Data Systems
516-682-8552

Paul Pechersky
Co-Chair, ISG Digital
Information
Interchange Group
E-Systems, Inc.
214-661-1000

Martin Plawsky
Co-Chair, ISG Design Integration Group
Grumman Corporation
516-577-1473

John Roche
Co-Chair, ISG Design Integration Group
McDonnell Aircraft Co.
314-777-7904

Jonathan Tilton
Co-Chair, ISG International Group
General Electric Co.
617-594-5492

Section Four: Planning Information Exchange within the Organization

Section Four papers are:

- *Engineering Data Management.*
- *A Flexible Manufacturing Technical Data Management System.*
- *Automating Electronics Manufacturing Documentation Management.*

A data management system incorporating Computer Integrated Manufacturing (CIM) system strategies is described by Sheth. This system has the ability to include data from different technical and business applications as well as that generated manually at diverse locations. Sheth summarizes an overall project's architecture and its benefits to give the reader another reference.

Beach and Jones discuss the data management system in place at IBM's Rochester, Minnesota plant, which won the Baldrige Quality Award. The Computer-Integrated Manufacturing system supports the flexible needs of this plant's manufacture of 3.5-inch disk drive products. Data management is a key component of an EIX system and warrants very careful attention.

Last in this section, Ginsberg describes manufacturing documentation characteristics and the associated effectiveness of various standards in this environment. This valuable standards comparison reduces the individual discovery efforts of many planners. These papers give the planner a set of internal considerations to be understood and designed into a quality EIX system.

Presented at the CASA/SME AUTOFACT '89 Conference, October 30 - November 2, 1989

Engineering Data Management

SHARAD SHETH
Electronic Data Systems Corporation

Corporations need a global data management solution to speed the Art-to-Part product cycle. This article describes the infrastructure of an enterprise data management system. This system is designed to control data generated by different technical and business applications as well as manual data. Technical applications include Computer-Aided Design, Computer-Aided Manufacturing, Computer-Aided Engineering, and Computer-Aided Testing. The control function covers applications running on geographically dispersed interconnected computers. The data management system also incorporates the corporation's information and project management practices and procedures. The concepts discussed are critical to a successful Computer-Integrated Manufacturing strategy and have been implemented and tested through currently available technology.

INTRODUCTION

Poor management and control of data leads to delays in transmitting information and reduces the effectiveness of that information to downstream users. Corporations need a global data management solution to speed the Art-to-Part product cycle. The solution must allow complete data control and ensure data accuracy and easy availability. This article describes a data management system that manages the many different types of corporate data according to the corporation's practices and procedures.

Engineering and manufacturing information takes many forms, each equally important to overall product development. Preparing, controlling, and distributing this information often requires labor-intensive manual methods. Designers commonly use manual methods and require supporting paper documentation and communication to convert designs into products.

Applications such as CAD/CAM increase the productivity of specific users but create islands of automation. These islands are not flexible enough to interface with existing systems.

Figure 1 illustrates a typical product life cycle. Each phase in the life cycle generates different forms of data. Some of this data passes through the subsequent phases of the process. For example, specifications defined in the Preliminary Design phase are passed to the Engineer and Build Prototype phase. Every corporation has practices and procedures that have evolved over time and govern the flow of product life cycle information.

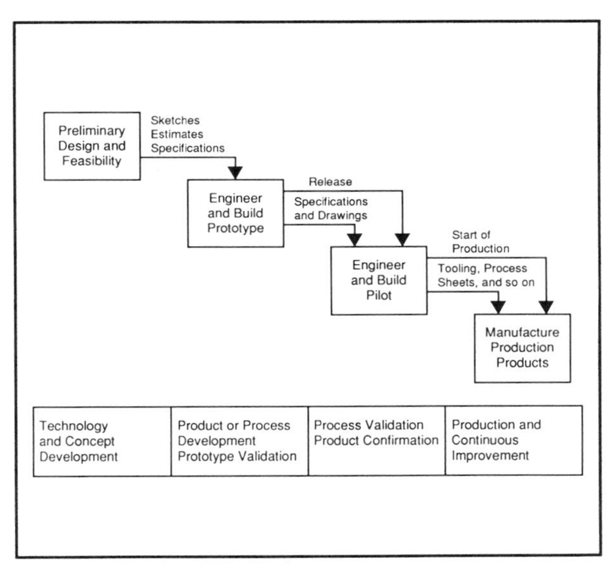

Figure 1. Product Life Cycle.

DATA MANAGEMENT ISSUES

Enterprise data, especially technical data, is a vital corporate resource. Corporations must manage this resource effectively to be successful, ensuring that it is shared, meaningful, and independent.

Automating the management, control, and distribution of product and manufacturing data is central to any CIM strategy. It provides the capability to optimize information flow within both the technical and office environments. The effort to integrate these environments can be minimized by using modular software design and standard hardware platforms.

Each advance in Computer-Aided Design, Manufacturing, and Engineering (CAD/CAM/CAE) speeds data creation, analysis, and modification. Therefore, data management systems must keep pace to fully benefit from these changes. Eliminating duplicate data and synchronizing data are essential—not only for engineering and manufacturing, but for finance and administration as well.

Present data management methods are inadequate to control the massive flows of information within corporations today. Users of information are frustrated by their inability to access the information they require. Further compounding this problem are the different organizational practices and procedures for controlling the information flow within a corporation.

Current Environment

Manually generating engineering drawings and other documentation to create or revise products leads to an enormous amount of paper. Even product development groups that are partly automated generate too much paper. Corporations have not yet automated and integrated four processes that computers can perform more efficiently on-line:

- Document and file cataloging.
- Document and file distribution.
- Engineering change tracking.
- Project management.

Engineers and plant-floor employees rely on blueprints, aperture cards, microfiche, and CAD system plots.

Distributing paper is labor and time intensive. For example, engineers revising products deliver paperwork to a reproduction department for distribution. If reproduction and distribution are inefficient, they hamper plant-floor operations. Information is obsolete when it arrives.

Many groups use CAD/CAM systems so incompatible that their data management problems are unique. They cannot move files efficiently between systems. For example, their CAM engineers cannot use on one system the CAD files they created on another system. Even their automated processes such as CAD/CAM provide no single point of control to give users good information.

Some computing platforms best support certain applications. However, unless organizations push to integrate heterogeneous systems, islands of automation will continue to cripple integration efforts.

Geographically dispersed businesses have added complications. The difficulties of combining dispersed product data to create a high quality product often reduce any advantages of locating in other states and countries.

Essential Features

A complete data management system must offer certain essential features. Separately these features could improve specific areas of data management; together they dramatically improve data management. The following paragraphs describe these essential features.

Users must have access to administrative and technical data through a single point of access from a variety of common terminals. The system must be able to use existing equipment as much as possible, reducing the need for special access hardware. The system also must use relational database technology to facilitate user-defined queries and reports.

The system must accommodate CAD/CAM data, raster files of converted manual drawings, and ASCII files. Unique file and document identifiers must ensure that the system displays the correct data on compatible hardware.

Converting all manual drawings into CAD drawings is prohibitively expensive, even though many must be quickly accessible. However, raster scanners can convert paper, aperture cards, microfilm, and microfiche into images that users can view on-line. Scanners are available to convert such drawings back into vector images so users can change them on-line, if necessary. Optical character recognition can in some cases convert raster documents into on-line documents for editing.

Users must be able to use system tools in heterogeneous and distributed environments. This portability minimizes the need for special equipment, encourages users to accept the system, and reduces system costs.

Organizations can identify required interfaces and

eliminate duplicate data by analyzing existing systems. Typical interfaces include CAD, bill of material (BOM), manufacturing resource planning (MRP), and shipping and receiving systems.

Users must be able to customize menu-driven display screens to conform to their own practices and procedures. System developers must thoroughly understand these procedures.

Where necessary, security constraints should limit a user's access to specific transactions and data files. The system should let users view, print, edit, and update files at specific phases in the product life cycle. These privileges can be made available through system utilities.

The system should include optical disk and Write Once, Read Many (WORM) technology. Through optical disk technology, corporations can cost effectively archive data. WORM disks can prevent data loss and provide an audit trail of data changes.

Through low-cost printers, the system should provide convenience prints for quick reference on the shop floor and in other departments. High quality hard copies can be produced on electrostatic plotters, and high volume needs can be satisfied using laser printers.

The system should reproduce digital images on paper, aperture cards, microfilm, and various optical and magnetic storage media. Viewing an image is often sufficient, and no reproduction is required.

Transmitting information over wide area networks (WANs) can satisfy the needs of plants and offices in geographically dispersed plants and facilities. Local area networks (LANs) can satisfy interdepartmental and plantwide needs.

Systems developers and suppliers must comply with applicable standards. Organizations must avoid proprietary standards to ensure that their systems are portable and capable of interfacing with other systems.

STRATEGY

Applications such as CAD, CAM, CAE, MRP, and BOM generally provide their own file management functions. However, bridges are necessary when applications work together. If one system changes, developers must upgrade software and bridges. Until now, this has been a time-consuming and unreliable process.

Figure 2 shows the integrated approach to data management. An integrated system of data management has several layers. The base physical layer consists of the different computing platforms used by the corporation. This layer is a mix of mainframes, midrange computers, and workstations. Data is stored in storage devices attached to these machines. The data is managed by Control Applications and defined by the four major functions analysts have identified as necessary for an integrated system: Document Management, Engineering Change, Data Distribution, and Project Management. The bottom layer comprises the various Engineering and Business applications used by the corporation. Some examples include CAD/CAM/CAE, MRP/BOM applications, and manual data that is scanned and stored electronically.

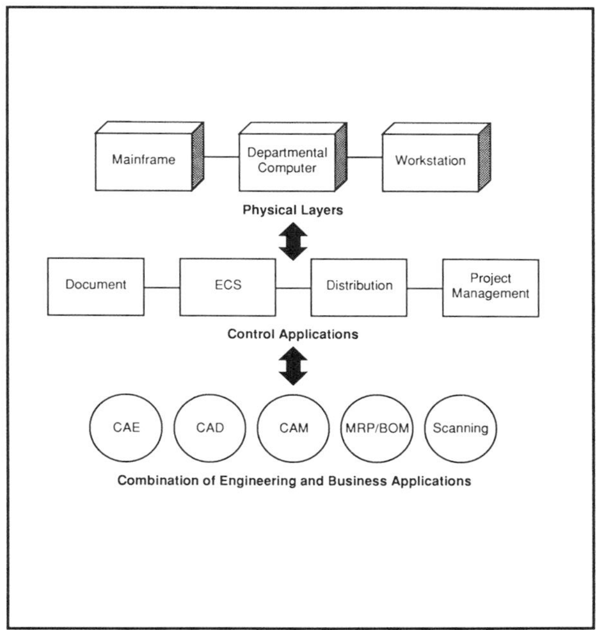

Figure 2. *Strategy.*

SOFTWARE FUNCTIONS

Analysts extensively examined user requirements to identify the four major functions that a total data management system must incorporate, as illustrated in Figure 2. These functions must provide features from which the user can choose. The following paragraphs describe the four major functions.

Document and File Management

Document and file management functions control all data access, including CAD/CAM files. This function is used most often by the engineering or manufacturing user. The more often they use it, the greater this function's benefits. Document and file management functions give users the following types of information:

- Description.
- Document number.
- Document size.
- Sheet number.
- Document class.
- Document form.
- Document status.
- Revision level.
- Superseding documents.
- Effectivity dates.
- Location.
- History.
- Distribution pattern.
- Relationships to engineering changes.

Users define and then activate or deactivate these attributes as their sites require. Access to each document is controlled based on the following features:

- User authorization.
- Document status.
- Revision level.
- Associativity.

These controls ensure users of greater integrity and higher confidence in system data.

Engineering Change Management

The engineering change management function controls the process of modifying engineering designs. This function tracks the change process from design initiation or revision through release and service, ensuring that everyone is notified of authorized changes. Users should receive the following kinds of information:

- Pending revisions.
- Change status.
- Responsible engineer.
- Authorizing activity.
- Effectivity by date and location.
- Affected documents.
- Item description.

Engineering change tracking and scheduling and electronic signoff are optional features.

Project Management

The project management function is the mechanism for organizing all engineering activities. It defines the events in the product life cycle, including completion dates, tasks, required resources, overlapping activities, and associated budgets. It then calculates the critical path for completing the project. Project management is the organizational tool for controlling the product development and implementation life cycle efficiently and cost effectively. Its four main components are as follows:

- Production readiness.
- Budget management.
- Resource management.
- Task and resource scheduling and tracking.

The project management function must interface with Document and File Management, Engineering Change Management, Document and File Distribution, and other key business systems to provide configuration management of a product through all stages of the life cycle.

Document and File Distribution

The document and file distribution function defines patterns for distributing information based on users' needs. The distribution network is unique for different user groups. After the network has been defined, automated or manual triggers activate distribution. Activation may depend on elapsed time or a project's completion.

For example, a trigger may start distribution for certain documents every six months; or a project manager may trigger a document's distribution after the completion of a project. A successful distribution function includes the following features:

- Multiple distribution patterns.
- Flexible distribution patterns.
- Acknowledgement functions.
- Intelligent routings.
- Wide area and local area distribution.
- Electronic distribution.
- Audit trails.
- Billing mechanisms.
- Multimedia data distribution.
  - -Magnetic and optical disks
  - -Aperture cards
  - -Paper
  - -Microfilm

Software Interfaces

Diverse computing platforms and information processing requirements lead to islands of automation that satisfy the requirements of specific areas. These islands of automation obstruct data sharing throughout the organization. A global indexing scheme that preserves individual applications must be provided to bridge between these islands of automation. Several systems typically require interface:

- BOM.
- CAD.
- CAM.
- MRP.
- Business systems.
- Office automation systems.

A detailed requirements analysis must precede implementation. This analysis determines both logical and physical interfaces to existing and planned systems. A successful data management system should interface with existing systems. Equally important, those who analyze requirements must be knowledgeable and experienced enough to recommend logical, cost-efficient solutions.

HARDWARE COMPONENTS

The hardware components of a successful system are connected through an open systems architecture. To provide many functions and capabilities, components should satisfy different requirements of performance and cost. Typical hardware components include the following:

- Input devices.
  - -Paper scanners in page and large format
  - -Aperture card scanners
  - -CAD files in vector and plot format
- Storage devices.
  - -Magnetic tape
  - -Magnetic disk
  - -Optical disk with Autochanger and drives
- Display devices.
  - -Engineering workstations
  - -Dedicated video display terminals (VDTs)
  - -Enhanced personal computers
- Output devices.
  - -Electrostatic printers
  - -Laser printers
  - -Dot-matrix printers
  - -Film printers
  - -Aperture card reproducers
- Processors.
  - -Mainframes
  - -Minicomputers
  - -File servers
  - -Engineering workstations
  - -Personal computers

The hardware architecture must be modular so that components and capabilities can be easily added and the processing power increased as required. Figure 3 describes the computing and communications infrastructure necessary to tie the hardware components together using a corporation-wide network. The network must allow peer-to-peer communication and

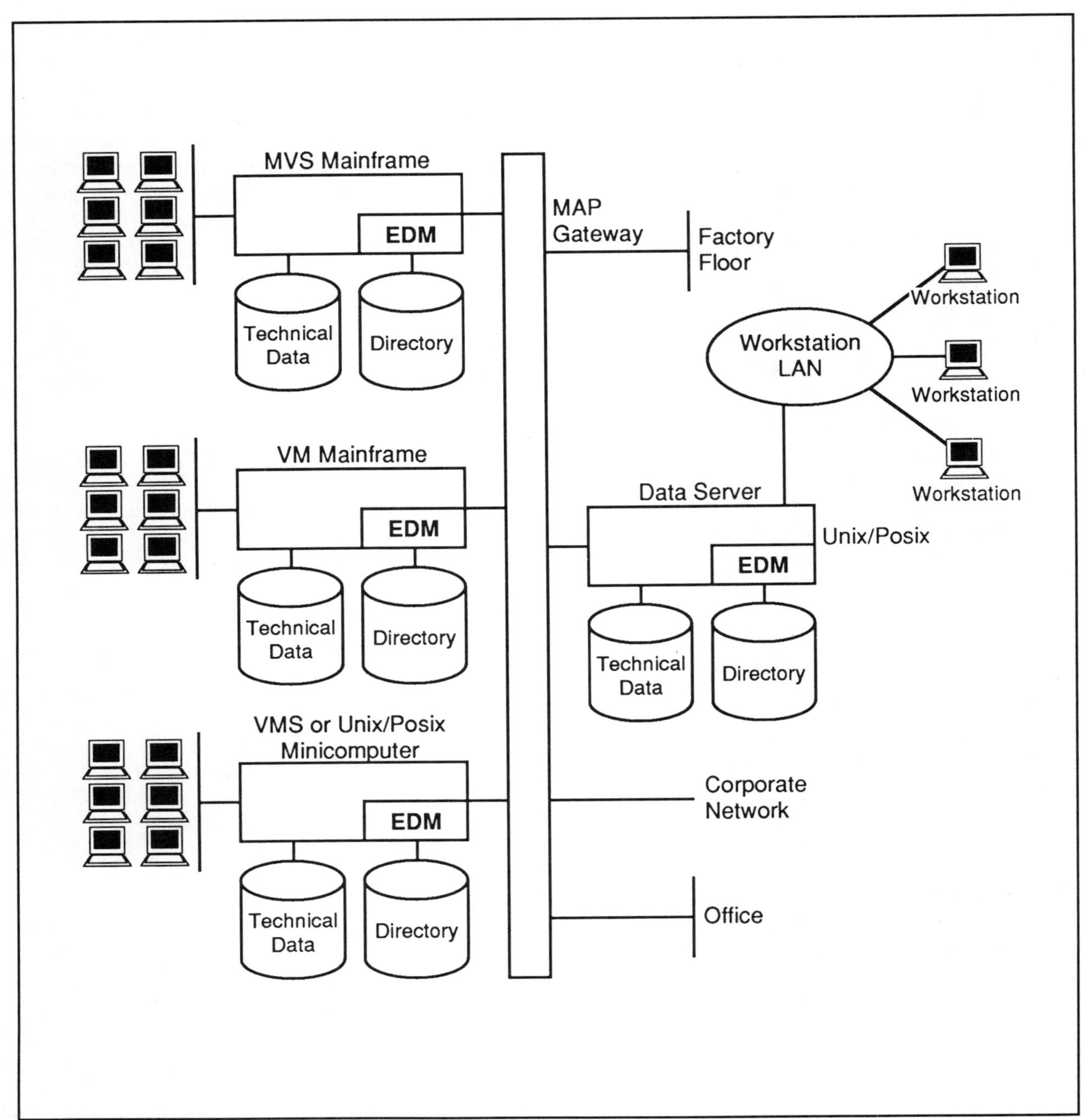

Figure 3. *Corporate Computing and Communications Infrastructure.*

between the different processors for fully distributed processing.

BENEFITS

An effective data management system provides global data management for better use of automation systems. Among the expected benefits are these:

- Increased productivity through improved information flow.
- Faster and more reliable access to data through a central data base.
- More effective management and control of information.

- Reduction in project lead times by ensuring that the right information gets to the right people at the right time.

- Reduction in distribution costs by eliminating couriers and mail delivery.

- Fewer product errors caused by unreliable data.

- Lower reproduction costs by reducing blueprints, microfilm copies, and photocopies.

- Smaller floor-space requirements by eliminating hardware such as blueprint machines and large storage cabinets for drawings.

- Elimination of departmental and shop floor "satellite files" by using a central database and by giving users data through strategically located terminals and printers.

- Elimination of document losses because of inefficient distribution.

- Less information redundancy.

IMPLEMENTATION RESULTS

The concepts described in the preceding sections have been implemented in part in several organizations. Several key points about data management can be stated as a result of these implementations:

- No turnkey solution can address all the data management needs of a corporation.

- The corporation must be amenable to changing its departmental practices and procedures to achieve effective data management.

- Stand alone image capture systems without document management are little better than an unorganized file cabinet.

- Proper planning and preparation are essential to successful implementation.

One division of a major electronics products manufacturer recently installed a system with some of the features described in this article. As in many other companies, this division relied heavily on paper for storing and transmitting engineering and manufacturing information. Valuable engineers' and professionals' time was wasted waiting in line at the file room for blueprints. Large volumes of original drawings had to be stored in file cabinets, which required increasing storage space. Drawing distribution was unreliable and inefficient. The division relied on aperture cards for user convenience and reduced storage and distribution costs; but data integrity was uncertain. Specifically, users complained of poor image quality, low revision accuracy, and often drawing unavailability—because they could not be located.

The system to solve this problem contained hardware components such as scanners, optical juke boxes, and several display terminals and printers. The heart of this system was document management software that manages engineering drawings, Engineering Change Orders, Parts lists, Design Standards, Tool Drawings, and Machine Repair manuals. The system has eliminated duplication and has decreased document access and distribution time for new and revised drawings. Mailing and filing drawings are now unnecessary.

In yet another case, a major transmission manufacturer needed to manage its CAD/CAM operations using data that was generated by different divisions and had to be shared with suppliers. The data was generated by different applications and resided on different computer systems. There was need for a notification and distribution system to reduce product development time. The data management system was required to secure data so that proprietary information was not released to unauthorized users. Also the practices and procedures of the different divisions and suppliers for managing their own information were implemented.

In this case the implementation of the data management system had to be preceded by a detailed study of the user and organization requirements. The implementation of the data management system was done in phases, to incrementally add functions.

The ultimate system will promote data sharing throughout a company. By sharing data during design, different manufacturers cooperating to make a product can manufacture consistent, high quality products and shorten their product-development cycle. However, manufacturers cannot expect to share data with each other until they streamline their own data sharing. They must have complete, accurate, and accessible data in their own systems, and they must ensure that all who use or manage the system are committed to its success.

CONCLUSION

All the discussion so far has focused on the obvious need for data management within a company that hopes to offer competitive products and services. The technology to implement data management systems exists and has been demonstrated for any application and corporation, no matter how large or small.

The primary force for ensuring the creation of a data management implementation plan is management's early commitment to the project. This commitment must be developed by corporations themselves. Outside vendors can only have limited influence without the commitment. Visionary people in the corporation who understand the organization's needs must generate this commitment and take the steps necessary to ensure its success.

Presented at the CASA/SME AUTOFACT '90 Conference, November 12-15, 1990

A Flexible Manufacturing Technical Data Management System

MARK J. BEACH, ALAN C. JONES
IBM Corporation

A data management system is in place at the IBM facility in Rochester, MN, that supports the collection, management, and analysis of production data for a 3.5-inch disk drive product. This system consists of computer and workstations connected through a local area network, and will support any serialized type of manufacturing. This data management system is an application of computer integrated manufacturing, providing an environment to manage the technical data needs for efficient manufacture of high quality products. The system is used in a manufacturing facility with dynamic production requirements. Its advantages include a flexible architecture to adapt to the needs of manufacturing.

INTRODUCTION

Over the past twelve years, the IBM facility in Rochester, Minnesota has been producing disk drive products. To efficiently manufacture such products, a comprehensive technical data collection system was necessary to collect parametric and pass/fail information from the many assembly stations and test cells located throughout the production facility. In the past, for each new product manufactured, a new technical data collection system was put in place. Because of the increased focus on shorter product cycle times and the high cost in time and manpower resources to develop and implement new technical data collection systems, a system has been developed that is flexible enough to satisfy the data management requirements of virtually any type of serialized assembly production. This paper describes that system.

Flexible Data Management System (FDMS) currently supports two autonomous 3.5-inch disk drive production lines in Rochester. FDMS facilitates the collection, management, and analysis of production data for these two disk drive products, and is an implementation of the evolving Computer Integrating Manufacturing (CIM) concepts. In FDMS, hardware and software are coupled to provide the personnel in manufacturing with a high quality, consistent system enabling them to use computers more effectively in their jobs. FDMS is comprised of a network of Personal Computers (PCs), Personal Computer AT™ (PC/AT) computers, and Personal System/2™ (PS/2) computers working in synergy to provide the data management functions necessary to assemble, test, and ship products of the highest quality.

SYSTEM REQUIREMENTS

When designing FDMS, the following criteria were used as guides:

Performance

FDMS collects and distributes data to both automated and operator attended workstations and testers. In some cases, data from the system is necessary before the operation can continue, therefore fast response is critical. In others, data is collected in a background mode for analysis and tracking, therefore fast response time is not critical. The goal is to respond to any critical request within an average of two seconds, and any other request fast enough to not impede the production line at that operation.

Availability

The manufacturing lines that FDMS supports are in operation 24 hours a day, at least five days a week. In addition, it is not unusual for production to work 40 of the 48 hours of the weekend. Given this criteria, FDMS is designed to run unattended for months at a time.

Recoverability

Availability of this system is critical to shipping the products that it supports. Architected into the system is a means of recovering from hardware or software problems with minimum production downtime. To date, the system has averaged less than one hour of unscheduled downtime per month.

Flexibility

FDMS has supported a manufacturing line from its very early stages of low production volumes through its current high production volumes. The manufacturing process and the product have gone through many changes during this time. To support these changes with the least amount of impact, flexibility is built into the system. This flexibility is built on the ability to dynamically create new data record formats through a system utility. Once this is accomplished, the workstation or tester can begin collecting the new data records as soon as they are ready.

Usability

FDMS supports user groups with high variety computer skill levels and expectations. First day workers to experienced computer scientists use this system daily. Consequently, the system must be adaptable to the varying abilities and needs of its users. This is accomplished through help text on demand, a consistent user interface, and text messages as opposed to cryptic numeric codes wherever possible.

Expandability

As stated earlier, FDMS has supported a manufacturing line from its very early stages to its current high volume capacity. It was obvious early on that the central database system would outgrow the capacity of a single PC or PS/2. With this in mind, the system is designed to be distributed across several computers, if necessary, to maintain the required performance.

SYSTEM ARCHITECTURE

The hardware used in FDMS is shown in Figure 1. The manufacturing floor is currently supported by a network of PCs and PS/2s using a token ring local area network (LAN). All manufacturing data collected is forwarded to the site Manufacturing Enterprise System for engineering reporting and analysis.

Processor Platforms

The only requirement for a node computer to communicate with the distributed Database PC is Netbios support through the token ring LAN. Nodes on the network include PCs, PC/XTs, PC/ATs, and various models of the PS/2 family. Communication to

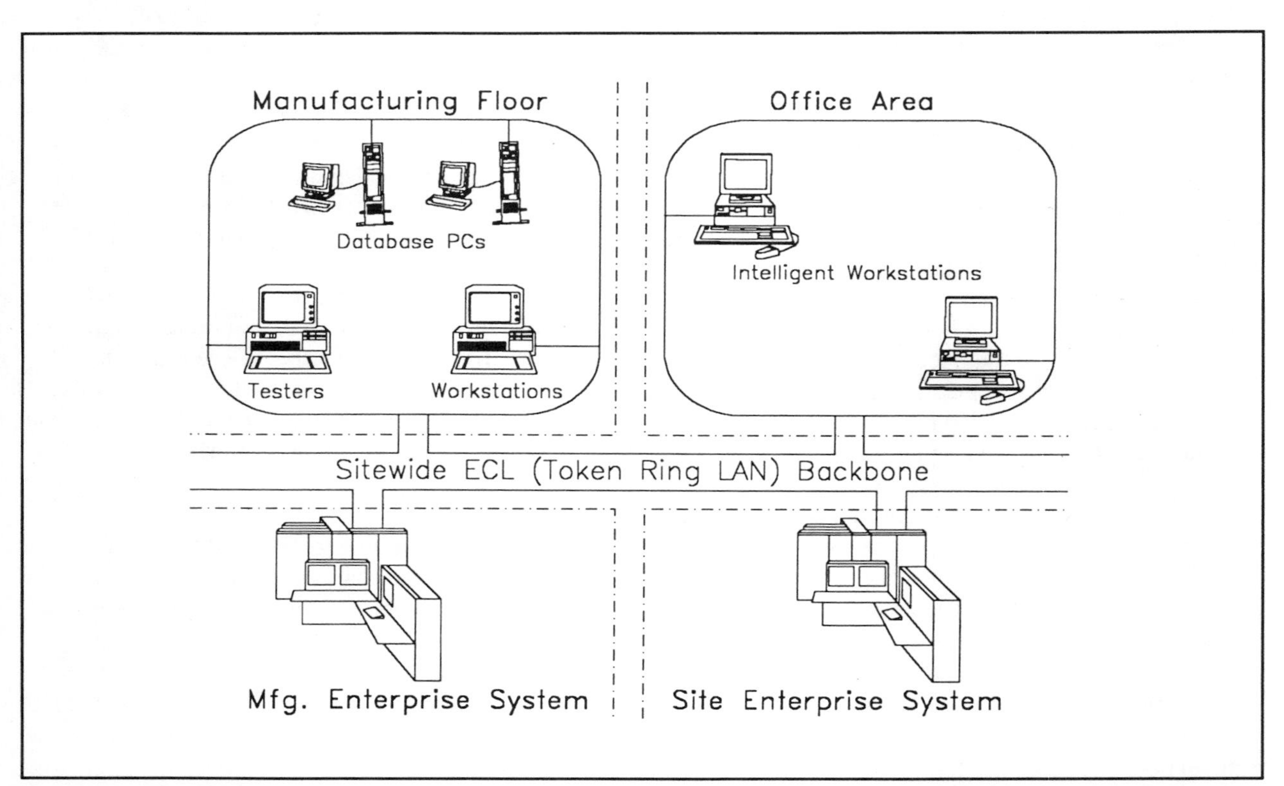

__Figure 1.__ Hardware and Local Area Network Layout of the Flexible Data Management System: Shown are the hardware components and the various accessibility paths to the production data from anywhere on site.

lower level devices, such as laser wand readers and Programmable Logic Controllers (PLCs) is through a subhosting concept, where a node computer on the network may support several of these devices. Communication to other families of computers is available through the IBM Distributed Automation Edition™ (DAE) product.

Local Area Network (LAN)

All communication between the distributed Database PC and the node computers is handled through the token ring LAN for the following reasons:

- The token ring LAN uses IBM's Establishment Communication Link (ECL). (The entire site where this production facility is housed is wired with ECL.)
- The token ring LAN allows the node computers to access the site's enterprise systems with no hardware modifications.
- Infrequent users of the system such as engineers or executives whose offices are remote from the production facility can access the system from an existing intelligent workstation with no additional hardware expense.

Other LAN types could be supported through the use of DAE.

Manufacturing Enterprise System

All the data collected in manufacturing is stored on the Database PC as well as on the manufacturing enterprise system. Functions that the enterprise system provide are the same as those that large systems are particularly well adapted to:

- Use of powerful analysis tools to accomplish in-depth interpretation of large volumes of accumulated test data.
- Archiving of old test data to tape for storage.
- Restoration of test data for file returns so that it can be analyzed to help identify and prevent the root cause of field failures.

Flexible Data Records (FDRs)

All production data collected from testers and operator workstations is in a flexible data record (FDR) format. In previous systems, these data records were in a predetermined form hard-coded into the program modules needing to process them. These modules included the tester control database, query, and report programs. In FDMS, every data record is associated with an "on-line" description of its contents called a FDR descriptor (FDRD). An example FDRD and FDR of the same type are shown in Figure 2.

The FDR and its FDRD are associated by two fields: type and version. The FDR type groups FDRs associated with a particular manufacturing step or process. For example, all FDRs with type "FT" could contain data from the disk drive final test station. The FDR version, along with its type, identifies the particular format of the data. As a data record changes over time, the version will identify the exact format for a particular record. The type and the version together uniquely identify a FDRD. For each field in a FDR the associated FDRD provides the following information:

- A 1-8 character name of the field.
- The length of the field, in bytes.
- The format of the data in the field. Valid formats include character, integer, floating point, hex and bit data.

Programs that process FDRs have the option of hard-coding the data formats as they have done in the past, or use the FDRD to locate the desired data by name. The latter method gives the processing program independence from changes that a data record may undergo over time.

Every FDR contains three sections:

- The first is the standard header. The standard header contains the FDRD type and version, component serial numbers, date and time pass/fail information, and other data fields that most testers collect. The standard header is the inflexible portion of a flexible data record. Other than the FDRD type and version, there is no system dictated need for a standard header; however, it allows for easier programming of applications to process the most widely used production data.
- The second section in a FDR is the parametric data section. This is a fixed length section of data mapped by the associated FDRD. The parametric data section format is completely free form but usually contains specific test results measure-

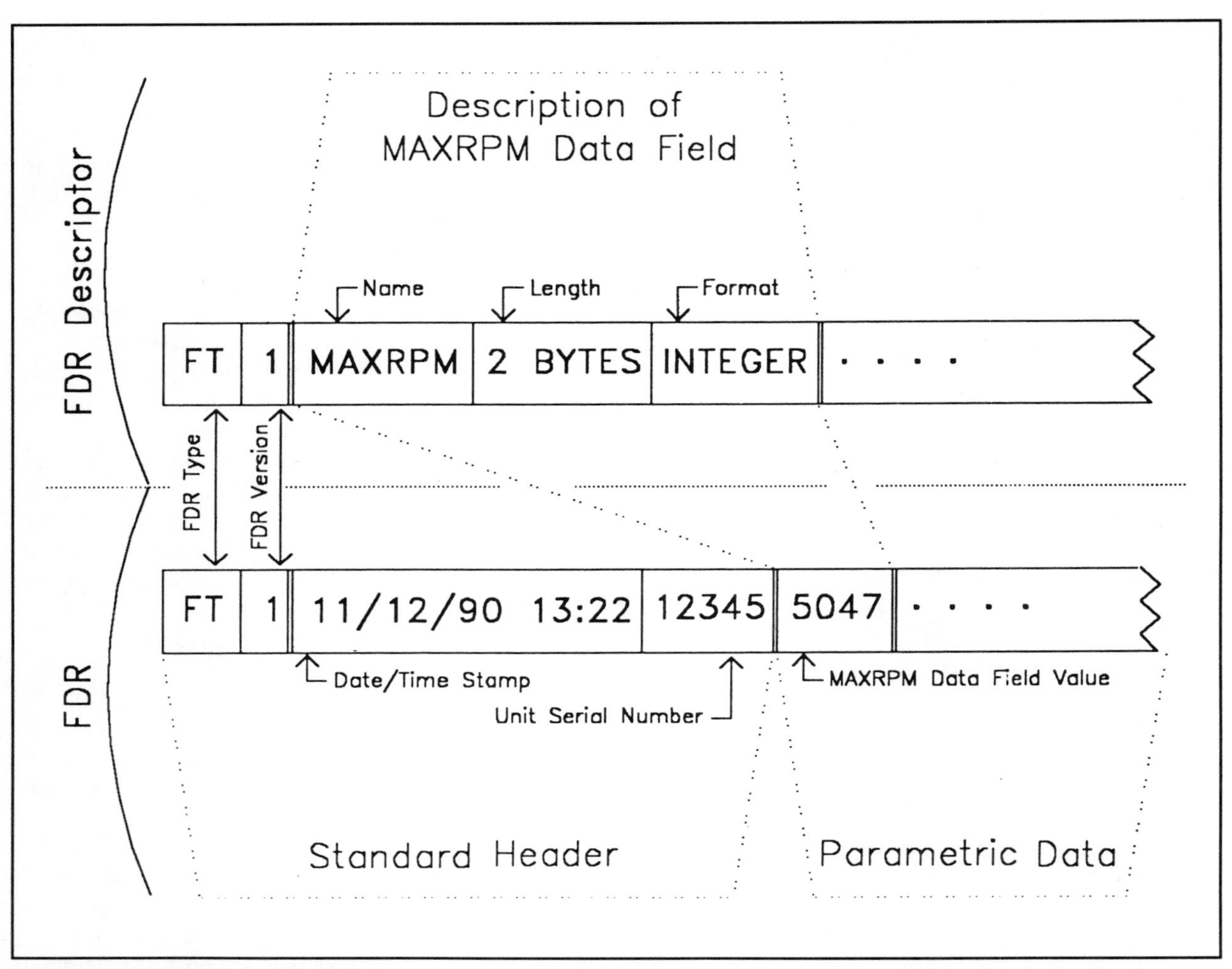

Figure 2. A sample FDRD and matching FDR: This shows the logical relationship between the attributes of a field in the Flexible Data Record Descriptor and an actual occurrence of the matching Flexible Data Record.

ments, and related information useful for the analysis of a failed disk drive.

- The third section in a FDR is the error section. This optional section is designed to accommodate repeating blocks of information to be collected. In disk drive manufacturing, an example of this would be the particular defects found on a disk surface. Since the count of defects can range from zero to thousands, this section allows for a variable amount of data while saving data storage space for those particular records with low defect counts. The format of the repeating block of data in the error section is also mapped by the associated FDRD.

Distributed Database Organization

As the FDRs arrive from the node computers, they are split into two database records for storage.

The first database record is the standard header. It is stored in the history database. It is a fixed-length relational database, and is keyed by six database keys: date/time, FDR type, and the serial numbers of the four components tracked for this product. Since each FDR contains a standard header, this database serves as an index to all data on the system.

The second database record contains the parametric data section and the error section of the FDR. This data is stored in separate databases, one database for each FDR type. These databases are keyed by the same date/time key that the history database uses.

The system guarantees uniqueness of date/time keys across any FDR type. Therefore, any history database record can be easily associated with its original variable length portion.

This database architecture may be distributed according to the following guidelines:

- As many computers as necessary can be used to hold the variable length databases, up to a maximum of one computer for each data record type.

- The history database can be housed on its own computer or can be part of the computer holding the variable length databases.

Distributed Database Query Capabilities

To retrieve data from the Database PC, several query transactions are supported. These query transactions allow the user to access any of the data that was previously collected. All initial queries to the system start with the specification of a criteria string, specifying values for the six database keys. Any combination of the database keys can be specified in the query criteria string. The response to this query is the collection of standard headers (history records) where the set of specified keys are exact matches. In the case of the date/time key, instead of an exact match, the specified value is treated as a starting point in time to begin the search. Once a node computer has retrieved these history records, the variable portions of any subset of the records can be retrieved by specifying the date/time key and the FDR type.

Real-Time Database Backup

To satisfy the availability and recoverability requirements for the system, a real-time continuous backup of any database is provided (see Figure 3). The backups are done at the database transaction level, thereby providing two varieties of protection:

- The obvious protection of hardware failure on the primary machine. If the primary machine fails, the backup machine can be brought on-line as the primary machine within minutes with no loss of data.

- To a lesser extent, software failure protection is also provided. Since the backups are accomplished at the database transaction level, software defects that corrupt the primary Database PC will not necessarily be executed on the backup Database PC. Defects in the Database PC application code above the database management level are not reflected to the backup in most cases.

Reporting and Analysis

There are two reporting and analysis capabilities of this system:

1. Retrieval and analysis of data for a single disk drive unit is handled by the query application program. The program is tailored to this type of analysis and is described in more detail later.

2. Analysis of data across more than one disk drive unit, with the exception of statistical process control, is handled at the enterprise system level. The data is available at the enterprise system level to be processed by the engineers using whatever tools best fit their needs. Yield reports, repair action effectiveness reports, and trend analysis reports using Pareto charts and histograms are examples of the types of analysis performed.

These analyses are essential to continually reduce defects in the manufacturing process. Since the data is forwarded to the manufacturing enterprise system in real time, engineers can analyze potential problems in a timely manner, and often correct problems before they become serious production inhibitors.

DISTRIBUTED APPLICATIONS

In this paper, the term "distributed applications" refers to any program that runs on a node computer and uses data retrieved from the Database PC to assist in accomplishing its function. The distributed applications discussed in detail in this paper are:

- Query. Allows operators to view any of the previously collected FDRs.

- Rework. Allows operators to enter repair actions that are collected and stored at the Database PC.

- Real Time Statistical Process Control (SPC) Generator. Automatically generates and tracks statistical process control charts from collected FDRs.

- Rework Expert System. Recommends repair actions for failed disk drives using an expert system shell.

- Prep to Ship. Assures that only the units passing all tests are shipped.

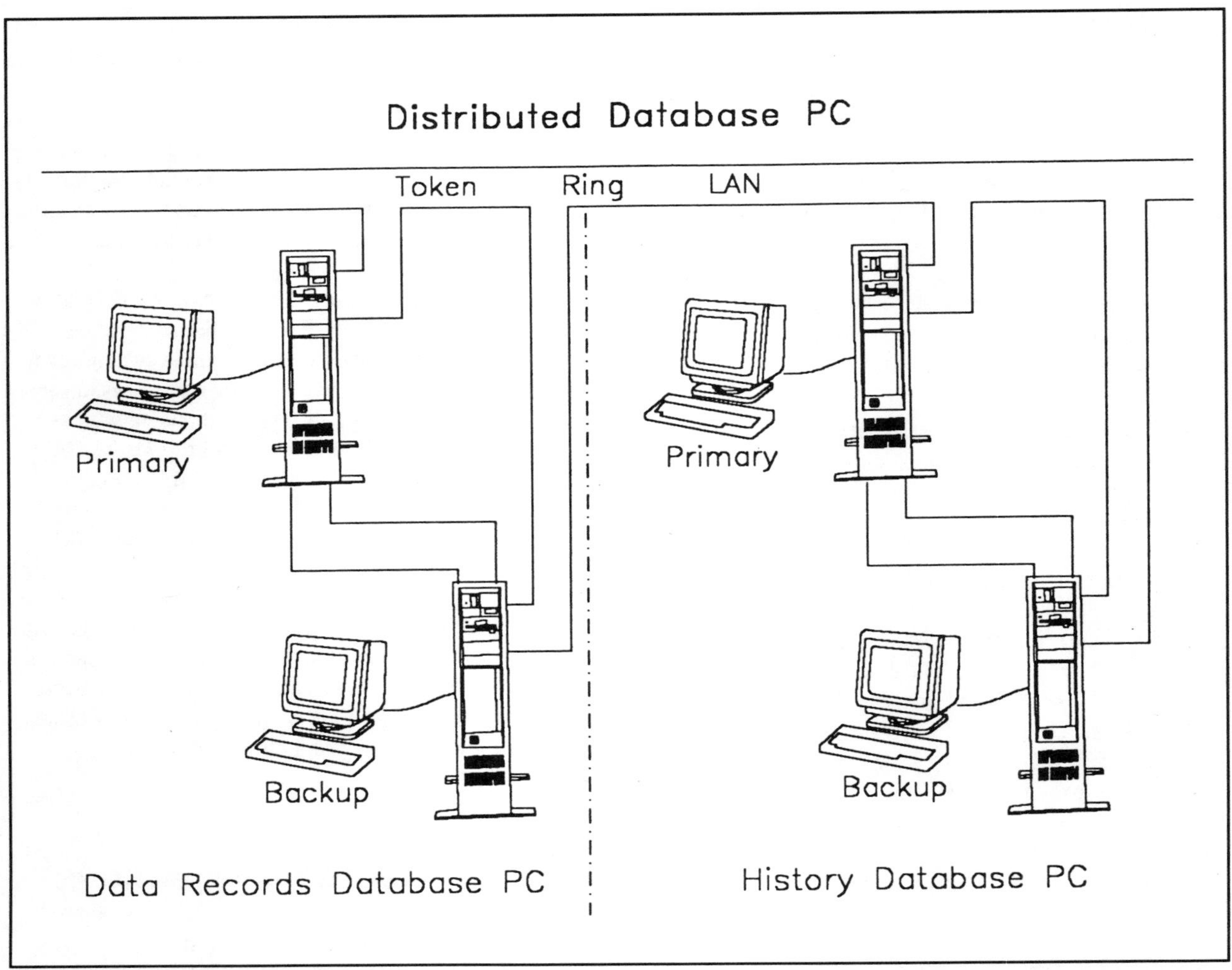

Figure 3. *Distributed Database PC Architecture: Shown are the primary backup Database PC architecture and physical split of the History and Data Records Database PCs.*

- Central Program Store. Assures that the software running on the various nodes is the correct level.

Query Application

Reduced paperwork in the manufacturing process has been identified as a key factor in reducing cost, reducing unit cycle time, and increasing the quality of the product. Each disk drive carries with it only a three-inch square piece of paper that contains the list of operations in the line. All other information is logged and tracked through FDMS. The query application is a program that will allow access to any data record on the system.

The query application allows an operator to use a barcode reader to wand in (or type in) any of the unit's component serial numbers and optionally qualify on the FDR type and date/time. The query application then dynamically constructs the query transaction and submits it to the Database PC. When the history records are retrieved they are displayed in chronological order for the operator to view. From this screen any record can be easily selected. Upon selection, again the appropriate query is dynamically constructed to retrieve the variable portion of the record and the associated FDRD from the Database PC. Upon receipt of this data, screens are dynamically formatted using the data definition information contained in the FDRD so that the data can be viewed in logical translation (integer, character, hex, etc.) and displayed with appropriate labels.

Rework Application

When a disk drive fails, it is sent to a rework analysis area. Here specially trained analyzers and engineers decide the action necessary to repair the unit.

A rework application program has been developed to assist the analyzers in logging suggested repair actions, viewing previously made repair actions, and logging the completion of previously suggested repair actions.

The rework application runs in conjunction with the previously discussed query application. When an analyzer has viewed the data using the query application and has determined the appropriate repair action required, this information can be entered on a screen. The information is then sent to the Database PC in the form of a rework FDR. Depending on the repair action necessary, one of the following will happen:

- If the analyzer decides to perform the repair action immediately, the rework program records both the suggested repair action and the fact that it has been completed.

- If the analyzer decides that the repair is best performed in another area, the disk drive is routed to that area. When the disk drive arrives at the repair station, the operator uses the rework program to retrieve the suggested repair action. When completed, the rework program records the completed repair action.

Real Time Statistical Process Control (SPC) Generator

The real time Statistical Process Control (SPC) generator application provides a complete system for tracking SPC parameters in a real time environment. In previous data management systems, SPC analysis of automatically collected parameters was accomplished by batch jobs. Since the batch jobs run only periodically, the analysis results were often too late to be of maximum benefit. This application performs ongoing SPC analysis as the data is collected, insuring that when a parameter goes out of control, notification (and therefore investigation and resolution) is immediate.

The SPC chart is defined and customized in a profile where the program user can specify the necessary parameters. These parameters include the operation and variable to be charted, the sample size, control limits, etc. The charts can also be tailored to flag out of control conditions depending on who can correct the process upsets. In addition, the profile contains a distribution list specifying who will be notified when a SPC chart goes out of control.

As FDRs are sent from the testers on the manufacturing line, the SPC station obtains the data in real time and creates a random sample to calculate the mean, standard deviation, and capability index. A graph of the most recent 25 means and standard deviations for each parameter can be viewed on the terminal screen.

Displaying the on-line SPC charts on remote PCs on the token ring LAN is through the PC LAN program. Remote PCs running the PC LAN program can access the SPC station in manufacturing and run a version of the SPC application to display and update the SPC charts.

When a SPC chart goes out of control, the display on the SPC station in manufacturing changes color to alert production personnel. In addition, a notice is sent through the site electronic mail system to the list of names specified in the profile.

Rework Expert System Application

As production volumes rose, so did the number of disk drive units that required rework analysis. Instead of simply training more rework analyzers, a rework expert system was written to make many of the decisions. This expert system's rules can make a suggested repair action on a high percentage of disk drives that have experienced a single test failure (where the analysis is limited to examining the previously collected data).

The rework expert system is used as a "screen" station where all disk drive failures are routed. If the expert system cannot determine the repair action from the data available at the Database PC, in conjunction with its rules, the failing disk drive is routed to rework analyzers. As the rework analyzers refine their knowledge regarding new failure types, the rework expert system rules are enhanced to handle these.

The rework expert system uses the same query transactions as the query application program, and the same FDR as the rework application program to record the repair action recommendations. Hence, it works in perfect synergy with the remainder of the manufacturing line involved with rework operations.

Prep to Ship Application

When a disk drive passes the final test in the process, its serial number is "wanded" into the prep to ship application. This application retrieves the history records for that serial number from the Database PC and verifies the following:

- The disk drive has passed all tests necessary to qualify to ship, and the tests were executed in the proper order.

- The disk drive did not participate in any experiments indicating it should be held for engineering analysis prior to shipping. A field in the standard header, the experiment ID field, is completed at any station where an experiment is being performed. Examples of experiments include line trials for new levels of parts and special tests that engineering needs to analyze before ship.

Central Program Store Application

In a high volume production facility there is more than one computer performing the same function. As the software on these computers is upgraded over time, it is difficult to ensure that each node is running the same level of software. In addition, releasing a new level of software to the production line can be a time-consuming and laborious task. The central program store application seeks to alleviate these problems.

This application enables the Database PC to act as a file server using the same network protocol as that to collect and retrieve FDRs. When a new level of control software is released to the production line, it is first stored at the Database PC. This software can be sent directly from the engineer's development workstation to the Database PC. Each computer needing the new software is then restarted. As part of the restart procedure, each node computer queries the Database PC to make sure that the level of software resident locally is correct. If it is not, the new software is automatically downloaded.

The microcode written onto each disk drive unit as part of the manufacturing process is also released with this mechanism. In the case of the microcode, the master copy of the microcode is automatically sent from the manufacturing enterprise system to the Database PC. As soon as the engineer authorizes the release of the microcode to manufacturing, it is ready.

Using the central program store application provides a single point of software distribution and level control, elements essential to producing consistently high quality products.

FUTURE DIRECTIONS

This system is still evolving. Major areas of current activity include:

- Implementation of an application known as condition notification. Here each FDR will be scanned as it arrives at the Database PC to look for defined trends in tester activity. Testers and engineers will be notified automatically when a condition is met, thus allowing problems to be discovered and corrected as quickly as possible.

- Further utilization of hardware independent CIM products will give the system increased flexibility. The ability to move an application from one node to another without program changes, a greater degree of database distribution, and hardware independent system design are desired benefits of this activity. Products currently targeted for further investigation and use include DAE, PlantWorks, OS/2™ and the OS/2 Database Manager, and Paperless Manufacturing Workplace.

- The rework expert system will be automatically invoked when a failure occurs. Currently the rework expert system takes action only when an operator "wands" in a disk drive serial number. As the rework expert system matures and the rules are expanded to make calls on the majority of the disk drive test failures, it will be able to do the analysis in unattended mode. The Database PC will notify the rework expert system of any test failures as soon as the failure occurs, and the analysis can take place immediately. The operator who is currently "wanding" in the disk drive serial numbers would be free to perform other tasks. Additional time savings would be realized by the rework analysis continuing unattended.

- A new Database PC function called Parts Database will be developed. This function will enhance the status and traceability of individual serialized components as well as fully assembled disk drives. In turn, the rework expert system application will use this additional information to investigate more complex parametric data interactions than is currently possible.

- The FDR system and query application will be enhanced to allow nonprogrammers greater control over the presentation of the FDR data. Allowing operators to configure the data on the screen into more meaningful formats, colors, etc. will increase the productivity of the person viewing the data.

CONCLUSION

Creating FDMS provided the following key benefits:

- The robust design of the system allows flexibility within a product's changing production environment. As the needs of manufacturing and engineering change, the system can be adapted to support those needs rather than being another stumbling block that must be dealt with in order to make the necessary changes.

- Savings in development cost of a new system each time we manufacture a new product. As FDMS was brought on-line to support its second product, less than 25% of the effort necessary to support the first product was expended.

- Manufacturing and engineering have benefitted from having one system supporting multiple products at the same site. Since FDMS is building on a common base and being reutilized on multiple products, it provides a consistent user interface and enhances functions on multiple product lines simultaneously. The data management system is not reinvented for each new product, resulting in a higher quality data management system and higher quality disk drives.

Reprinted from ELECTRONICS MANUFACTURING ENGINEERING,
Second Quarter 1992, Vol. 7, No. 2

Automating Electronics Manufacturing Documentation Management

GERALD GINSBERG, PE,
Component Data Associates Inc.

Manufacturing engineering involves creating and maintaining lots of documentation--bills of material, fabrication and assembly procedures, artwork for photomasks, drilling tapes, test procedures, engineering change orders, and so on. In today's highly competitive environment—with greater product diversity, more use of subcontractors and teaming, shorter product lives, and the need to move products among factories—the task of managing documentation is becoming much more demanding.

Also, benefits to the electronics manufacturer of managing documentation efficiently are significant. In fact, many companies now realize automating the documentation process is just as important as automating the manufacturing process.

This article describes recent advances in computer-aided design (CAD) and computer-aided engineering (CAE), relating them to the documentation needed by manufacturing engineers. The information will help you as a member of a concurrent engineering team, as a purchaser of contract services from assembly houses, and as a member of the customer service team supporting a major OEM customer.

Crossing the Design Bridge

While logic and circuit simulators, layout editors, test generators, and placement and routing tools are all part of the automation process, they remain individual islands of progress. Standard data formats that will allow data to pass among these pieces are only now being developed, and tools to manage and integrate the volumes of data are largely neglected.

The primary problem facing electronics manufacturers is the tremendous amount of documentation that must be created during the design process. The solution is to stream-line the design and documentation of advanced digital systems. Existing hardware description languages are incapable of accomplishing this because their evolution was not planned.

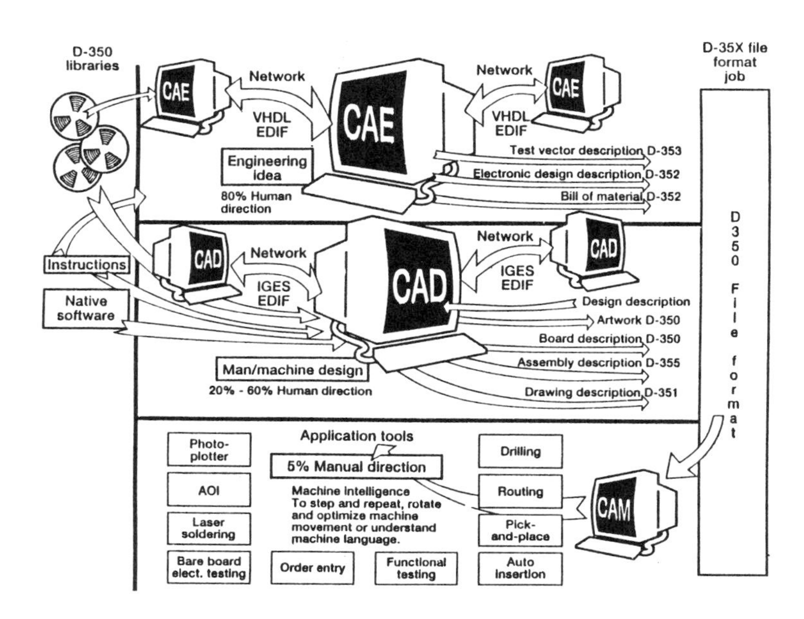

The need for hardware description language to input design data became apparent by 1979. This language would provide a human readable design description, support communication of design data among vendors (second sourcing) and between vendors and users (design specifications), integrate the activities of designers and manufacturing engineers working at different levels of abstraction, and increase the reusability of hardware designs and descriptions.

In addition to a hardware description language, an intermediate format was required to interface between standards and languages. Thus, several standard languages were developed that you can use to describe various automation functions, including printed circuit boards, phototooling, or information related to these topics.

Some of these languages have not necessarily been developed for pc board functions, but because of their applicability to other facets of the design process, they can also be used in pc board applications. These include such standards as IPC-D-35X, IGES/PDES, VHDL, and EDIF. In addition, users may have their own native data format standards related to the equipment they purchased.

Some companies have standardized within the various departments within their organization to allow electronic communication of product data. These are usually proprietary formats, not industry standards.

Such digital descriptions are desired to facilitate the automation process for producing parts. On-line electronic data transfer is possible and intended to end the need for human intervention, facilitate storage of documentation in a format other than paper (archiving), and create a standard means of describing design data that can be transported to machines other than the one on which it was created.

IPC-D-35X Series

The Institute for Interconnecting and Packaging Electronic Equipment (IPC) has generated a set of companion documents, the IPC-D-35X series, that consists of the following:

- ANSI/IPC-350, "Printed Board Description in Numeric 20 Form."

- ANSI/IPC-D-351, "Printed Board Drawings in Digital Form."

- ANSI/IPC-D-352, "Electronic Design Data Description for Printed Boards in Digital Form."

- ANSI/IPC-D-353, "Automatic Test Information Description in Digital Form."

- ANSI/IPC-D-354, "Library Format Description for Printed Circuit Boards in Numeric Form."

- IPC-DG-358, "Guide for Digital Descriptions of Printed Circuit Board and Phototool Usage per IPC-D-350."

- ANSI/IPC-NC-349, "Computer Numerical Control Formatting for Drillers and Routers."

The IPC-D-35X series was developed to specify pc board data in a machine independent digital format for communication from design to production and encompassing fabrication, documentation, assembly, and testing. The format supports data communication among computer-aided engineering (CAE), design (CAD) and manufacturing (CAM) systems.

The IPC series of standards starts with IPC-D-350, which describes the fabrication and artwork of a pc board. It has grown into a series of standards that describe the pc board many different ways including the schematic diagram, assembly drawing, electrical description, test data, etc. These standards also have the potential of describing the systems in which the boards reside.

The IPC-D-35X series is flexible as it allows many options. The structure of the standards allows adding new information and concepts by developing new data information modules that describe a particular parameter or facet of the design. As future enhancements are required, the concepts used in the IPC-D-35X series can have record formats added to assist in describing parameters needed by the design community.

IGES/PDES

The Initial Graphics Exchange Specification (IGES) is a communication file structure for data produced on and used by CAD/CAM systems. It serves as a receptacle for the data generated by commercially available interactive graphics design drafting systems. This structure provides a common basis for the automation interface.

IGES 1.0 was published as a National Bureau of Standards report in January, 1980 and was approved as ANSI standard Y14.26M in September, 1981. The

Design Language Process Functional Matrix

	System				Box				Board				Component			
	IGES	EDIF	VHDL	IPC	IGES	EDIF	VHDL	IPC	IGES	EDIF	VHDL	IPC	IGES	EDIF	VHDL	IPC
Behavioral Description																
1. General			●				●					●		●		
2. Quality Level			●				●					●		●		
3. Signal			●				●				●				●	
4. Ports			●				●				●				●	
5. Quantitative performance			●				●				●				●	
6. Operating range			●				●				●				●	
7. Safety			●				●					●		●		
8. Simulation—behavioral			●				●				●				●	
Functional Description																
9. Functional partitioning	●	●	●		●	●	●		●	●	●		●	●	●	
10. Form factor	●				●				●				●			
11. Algorithmic description			●				●				●				●	
12. Interface control & limit	●				●							●		●		
13. Environmental test parameters	●				●							●		●		
14. Simulation & functional			●				●				●				●	
Logical Description (Digital)																
15. Symbol definition		●		●		●		●		●		●		●		●
16. Signal		●	●			●	●			●	●			●	●	
17. Ports		●	●			●	●			●	●			●	●	
18. Timing (description)		●	●			●	●			●	●			●	●	
19. Simulation—logic (including models)		●	●			●	●			●	●			●	●	
Circuit Definition (Analog)																
20. Symbols		●		●		●		●		●		●		●		●
21. Gain charts/V-I plots		●				●				●				●		
22. Frequency plots		●				●				●				●		
23. Propagation delays		●				●				●				●		
24. Timing description		●				●				●				●		
25. Quantitative performance		●				●				●				●		
26. Operating range		●				●				●				●		
27. Q, R, & M calculations		●			●				●				●			
28. Simulation (circuit)		●				●				●				●		
Simulation																
29. Fault Simulation			●				●				●				●	
30. Test Vectors			●				●				●				●	
31. Thermal	●				●				●				●			
32. Vibration	●				●				●				●			
Net List																
33. Design Rules		●				●				●				●		
34. Parts		●	●	●		●	●	●		●	●	●		●	●	●
35. Interconnectivity		●	●	●		●	●	●		●	●	●		●	●	●
Physical Design Layout																
36. Physical design rules	●				●							●		●		
37. Dimensions/tolerances	●				●				●				●			
38. Package interfaces	●				●							●		●		
39. Material properties	●				●									●		
40. Reference designators				●				●				●				●
41. Cabling conductors	●				●							●		●		
42. Detailed thermal analysis																
43. Detailed R&M analysis																

Design Language Process Functional Matrix

	System				Box				Board				Component			
	IGES	EDIF	VHDL	IPC	IGES	EDIF	VHDL	IPC	IGES	EDIF	VHDL	IPC	IGES	EDIF	VHDL	IPC
Physical Documentation																
44. Detail/package drawings	●				●							●		●		
45. Reference designators	●				●							●		●		
46. Dimensions and tolerances	●				●				●				●			
47. Material construction	●				●							●		●		
48. Assembly drawing and notes	●				●							●				
49. Parts list	●				●							●		●		
50. Fixturing	●				●				●				●			
51. Pattern geometry	●				●							●		●		
52. NC data	●				●				●				●			
Assembly & Test																
53. Assembly specification		●				●				●				●		
54. Test/burn-in requirements		●				●				●				●		
55. Other Q, R, & M testing						●				●				●		
Installation																
56. Drawings				●				●				●				●
57. Tech manuals				●				●				●				●
58. Shipping container drawings				●				●				●				●

most recent version of the standard (Version 3.000 published in April, 1986) is ANSI Standard Y14-26m-1987.

The complexity of the IGES project is enormous due in part to the need for building in flexibility for future growth. To handle the complexity, future specifications will be developed using a methodology that divides the problem into three parts: logical, conceptual, and physical.

The methodology for data storage is an entity attribute database. Each record in the file associated with an entity contains parametric data related to that type of entity. Any entities that have a dependency relationship with other entities have within their records a pointer defining that relationship.

Extensions of IGES will include representation of the following design-related information: Integrated circuit design and connectivity, testing, simulation, inspection, and hierarchial electronics system design.

The IGES Group is also involved with developing the Product Data Exchange Specification (PDES). This effort addresses a different technology base than IGES. IGES is for information exchange between databases that must be interpreted by humans. PDES is a complete database exchange that does not require human interpretation.

PDES is the U.S. contribution to the international STEP (Standard for the Exchange of Product Data) development effort. The goal of the international effort is to develop a single standard for complete product data exchange.

VHDL

VHDL—the "Very High Speed Integrated Circuit" (VHSIC) Hardware Description Language—was approved as an IEEE standard (1076) in late 1987 and mandated by the Department of Defense under MIL-STD-454 Revision L for application-specific integrated circuits (ASICs) in early 1988. As such, it makes a significant impact on both IC and pc board design.

The specific purposes of VHDL are to provide a standard medium of communication for hardware design data, represent information from diverse hardware application areas, support the design and documentation of hardware, and support the entire hardware life cycle. A tenet of VHDL is the design specification should remain independent of its implementation. This pivotal rule focuses on the reusability of parts and ease of maintenance. The dedicated emphasis on commonality and reusability is structured to reduce the maintenance costs and effort demanded of manufacturers when parts are unavailable or when there are design revisions.

VHDL is not the easiest design language to learn and apply for any specific function. Thus, there are instances in which another language may be more appropriate. For most designs, however, VHDL permits replacing proprietary languages and applying top-down methodologies. It results in a system design emphasizing behavioral aspects of a circuit or pc board.

The major portions of a VHDL description are entities, architectures, and configurations. An entity is a design block or component and, at its lowest level, could be a single Boolean gate.

An entity declaration includes the number, type, and direction for each type of pin along with a series of parameters (called "Generics") passed to the component model (called the "architecture"). Entities can be thought of as schematic symbols; architecture as the simulation model for a symbol; and configurations as the link associating the entity with one or several different architectures.

The packaging facilities inherent in the VHDL language, along with the features permiting abstraction of data, allow unambiguously specifying the way a design should function. Thus, the VHDL simulator provides very high quality feedback, optimizes design efficacy, and advances the usability of CAE tools for hardware designers.

EDIF

The Electronic Design Interchange Format (EDIF) facilitates movement of electronic design data between sophisticated databases in a commercial environment. This provides links between disparate systems as well as corporate control over information content. EDIF is intended to represent IC and electronic design data, not mechanical design data.

EDIF can represent the electrical characteristics necessary for implementing electronic products. It can express library/cell organization; provide extensive data/version control; describe the cell interface, cell details (contents), and technologies; and represent timing, geometry, and physical objects.

EDIF addresses the IC description necessary for

fabrication. This includes modeling and behavioral aspects and some pc board descriptive material.

A net list generally describes ports or pins of parts and their interconnections. EDIF net lists may also contain system or user-defined properties. The properties may indicate electrical parameters such as timing, power rating, tolerance, capacitance, and loading. Net lists are commonly used to link a logical design to a physical layout.

An EDIF schematic includes all of the information in the net list. In addition, it provides a graphical representation of that data. A schematic diagram is often part of the final documentation for an electronic system.

The EDIF printed circuit layout view is intended to include all of the physical design information for a circuit board, including traces, lands, targets, holes, planes, and vias. In addition, it incorporates all of the part and connectivity data that can be specified in the net list view.

CALS

The Computer-aided Acquisition and Logistical Support (CALS) program is a Department of Defense initiative to accelerate the use of digital technical information for product acquisition, design, manufacture, and support. The benefits expected include the following:

- Reduced acquisition and support costs,
- Elimination of duplicative, manual, error-prone processes,
- Improved reliability and maintainability of designs directly coupled to CAD/CAE processes and databases,
- Improved responsiveness of the industrial base (i.e., to be able to rapidly increase production rates or sources of hardware based on digital product descriptions).

As a result of CALS, the various design languages have been rated according to their capability for handling electrical product data descriptions and recommendations have been made as to how to implement and use these languages (see chart).

Section Five: Planning Information Exchange External to the Organization

Section Five papers are:

- *Electronic Interchange of Product Definition Data Between Companies.*
- *EDI From a Supplier's Viewpoint.*
- *The Success of Customer/Supplier Information Exchange.*

A communication and translation system for technical data interchange installed at two manufacturing companies is described by Carringer. Quality Function Deployment was used in the requirements definition involving concurrent product and process development. This process enables effective electronic interchange of CAE/CAD/CAM specifications.

Foote then discusses the EDI program at E.I. Du Pont de Nemours and Company. Background, implementation, and standards issues and resolutions are included.

Finally in this section, the use of EDI to assist customers and suppliers is discussed by Johnston. Benefits are discussed with examples of specific corporations successfully using this technology. This brief look at information exchange with external business partners provides an initial understanding for EIX system planning activities.

Presented at the CASA/SME AUTOFACT '89 Conference, October 30 - November 2, 1989

Electronic Interchange of Product Definition Data Between Companies

ROBERT A. CARRINGER, CMfgE
International TechneGroup, Incorporated

The electronic interchange of product definition data (CAE/CAD/CAM) between companies supports the goals of concurrent product/process development. The requirements for this interchange of technical data have been identified using Quality Function Deployment. A communication and translation system has been developed which addresses these requirements and has been installed at two manufacturing companies. This paper describes those requirements, an approach for implementation, the two installations, and lessons learned.

INTRODUCTION

Most manufacturing companies aspire to achieve world class competitiveness and are in the process of implementing technology and cultural changes to achieve it. They strive to improve product quality, shorten product development time, and reduce costs.

Trends witnessed in the automotive and general industry segments indicate similar programs and activities across the manufacturing business. Manufacturers obtain their suppliers' technical expertise in the design of subcontracted components. The trend is away from a "build-to-print" operation to a distributed engineering environment. (1) Distributed engineering is implemented more easily within the supplier community with the use of new technology for product design, evaluation, and manufacture. Suppliers of all sizes are setting a trend for installing (and using) CAE/CAD/CAM technology. Suppliers are asked to participate with the manufacturer very early in the product life cycle, which requires the exchange of product definition data between the manufacturer and supplier.

The product definition data or design data (2), contains geometry, nonshape information and text annotation. In the past, it has traditionally been exchanged on engineering drawings. The drawings have been sent via the postal service or express mail services on paper, mylar or magnetic tape (in electronic form).

To reduce the time consumed in repeatedly shipping product definition data to suppliers, manufacturers have begun to transmit electronic files containing the data. The files, usually in the Initial Graphics Exchange Specification (IGES) format (3), transfer between different CAE/CAD/CAM systems with translators. Improvements in the cost and speed of electronic communications enable cost-effective interchange of this data. This activity is defined as Technical Data Interchange (TDI).

The purpose of this paper is to:

- Describe the requirements of electronic Technical Data Interchange.
- Present an approach for implementation based on case study experience.
- Summarize the lessons learned, benefits and challenges of Technical Data Interchange.

REQUIREMENTS FOR ELECTRONIC INTERCHANGE OF PRODUCT DEFINITION DATA

Working with a major manufacturer of diesel engines, ITI developed and documented the requirements for technical data interchange using the Quality Function Deployment (QFD) methodology. (4) QFD, used frequently by the automotive industry, is a product development methodology used to cross-reference all product or service engineering characteristics to a customer driven need. The customer needs and engineering characteristics matrix form the core of the House of Quality as shown in Figure l. All customer needs, the voice of the customer, were

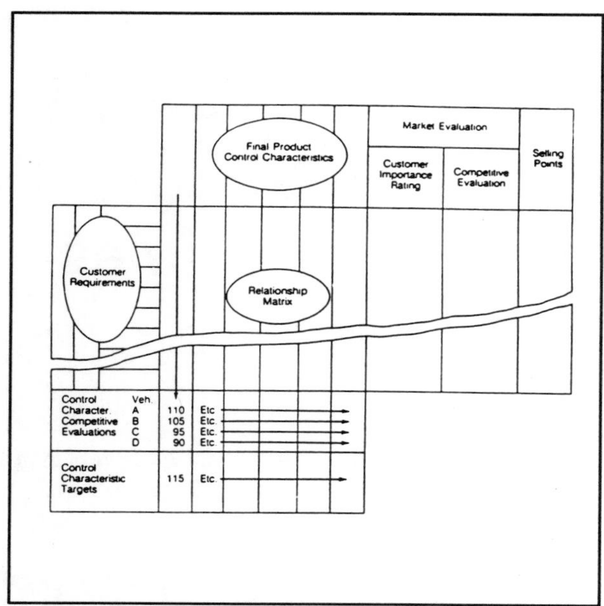

Figure 1*. House of quality example. (5)*

developed during the June to August 1989 time period.

To identify the voice of the customer, the project team interviewed several personnel from many organizations at the manufacturer's and suppliers' sites. All potential users were considered customers of the system; they were categorized as Engineering, Purchasing and Supplier. Engineering covered personnel with interests ranging from design, manufacturing liaison, product development and customer liaison. Purchasing customers included both purchasing agents and technical support engineers. Suppliers involved in the projects typically were smaller companies than the manufacturer and, therefore, had personnel who participated in several roles. In general, they could be grouped in similar categories as the manufacturer's personnel, but were cumulatively grouped into one category.

Customer needs, gathered during interviews with potential users, were later consolidated and ranked by the customers. Alternate solutions to the needs were analyzed to identify existing perceptions by customers. The customer needs are shown in Figure 2 and some are further described in subsequent paragraphs.

Measurable Benefits

The engineering and purchasing customers wanted the system to provide measurable benefits. This need was rated more important than needs like "cheap to install and use." The benefits expected by the users were reduced lead time and reduced reengineering by suppliers. Engineering indicated that this need was not fulfilled by notional data, but could only be satisfied with actual data collected during pilot programs.

3-D Data

Suppliers frequently requested that complete part model geometry be transferred to reduce model clean-up, reengineering or complete reconstruction. (Limitations in CAD/CAM systems and IGES translators seemed to be the main cause for this requirement to be emphasized by participants.) Suppliers asked for both model and drawing data; drawing data was most important.

Accurate IGES Translations

Accurate and reliable IGES translation software was identified by all customers as an important requirement. This meant obtaining near complete translation. The customers wanted to know what data could not translate and what could but without 100% accuracy.

Fast Communications

Customers wanted to transfer the IGES files to the network quickly; 4800 bpi asynchronous speeds appeared to be satisfactory. In addition to telecommunications speeds, "fast communications" meant that the process of using the software system was quick with a minimum number of steps. An alternate approach to fast communications was to enable the data transfer process to operate in the background mode.

Domestic/International Access

Purchasing wanted the telecommunications portion of the system to have complete domestic reach and near-complete, free-world international reach. The network needed to be accessible to suppliers in the U.S. through local telephone networks. Internationally, Purchasing wanted to be able to send/receive design data to suppliers easily (without special actions needed to address country-specific issues). With the focus on obtaining components from the world's low cost, high quality supplier, Purchasing ranked this need third on their priority list.

Data Security

Engineering and Purchasing expressed the need for data security with respect to sending electronic files through networks. They wanted to limit the supplier's access to data in only those files the manufacturer chooses, to insure the files are not sent to the wrong supplier, and to make sure files sent do not contain information that would compromise the manufacturer's proprietary rights or competitive edge..

		1	2	3	4	5	6	7	8	9	10	11	12	13	14	15	16	17	18	19	20	21	22	23	24
Insures Data Security	1																								
Simple and Easy to Use	2		○	○	○			◎		◎	△											△			
Automate Data Transfer	3		◎	◎	◎			◎		◎	◎														
Fast Data Communication	4						◎	○		◎	◎		◎	◎								○			
Background Operation	5							◎		◎	◎														
Reliable IGES Exchange	6															◎	◎	◎	◎	◎					
Capable of 3-D Data	7															△	△	◎	◎	○					
Apollo Resident System	8		○	○	△			○		○															
Send Descriptions in File	9																								
Reduces Lead Time	10							△		○	○		○	○				○	○	◎			○		
Purchasing Interface to S	11																								
USA/International Access	12																					○	◎		
Customer support for new	13																								
		1	2	3	4	5	6	7	8	9	10	11	12	13	14	15	16	17	18	19	20	21	22	23	24

Figure 2. Customer needs for TDI.

Simple and Easy to Use

Engineering wanted the communication system to be simple to learn and easy to use. They wanted the user interface to be intuitive, similar to existing engineering systems and accessible from their engineering workstation. They wanted the process of sending and receiving files to be fast with few steps and limited data to enter. The user interface needed to be like a CAD environment with mouse interaction, icons, and other interaction types common to CAD systems.

APPROACH FOR IMPLEMENTATION

Components of Technical Data Interchange Capability

Transmission media, a critical component of TDI, provides the physical transfer of the data as a basic service. While several alternatives exist for the transmission media, a public network, also called a value-added network, provides many features above the basic service that benefit a TDI program. A public network typically provides local access throughout its reach. It provides an insulation function by keeping the supplier from connecting into the manufacturer's computer systems. Using a public network, the manufacturer is relieved from the logistics of developing and supporting national and international communication capability. The public network also provides several layers of data security and authentification control. Usually, they have high uptime and are reliable in operation. When compared to the alternatives of internal networks and direct dial, the public network makes a TDI program simpler to implement and support.

Data translation capability, an important component of TDI, provides the means for CAE/CAD/CAM systems to exchange product definition data. The CAE/CAD/CAM systems at different manufacturing sites are usually different brands and, therefore, cannot exchange data directly. The alternatives for exchanging the data are direct or neutral translators. For companies exchanging data with several suppliers, the typical choice is a neutral format. The most common neutral format used is the national standard, IGES. The Initial Graphics Exchange Specification (IGES) is supported by major system vendors in varying degrees of implementation. Its successful use requires significant implementation effort which often stymies casual users. The data translation must be accurate. It must provide a mechanism to exchange the data to the other system in a form functionally equivalent to the source data. It must also be reliable so the suppliers do not need to recheck the translation for every exchange.

Another component of TDI is the gateway connection to the network. Each user must have their workstation/terminal system connected to a telecommunications capability to reach the network. For most engineering users, the environment is an engineering workstation. Whatever the environment, the engineer needs to invoke the communications to transfer a file by using their CAD system. For casual users, a less powerful, less expensive solution can be obtained through a personal computer connection. In both situations, the users need a software system that provides communications, compression, encryption and administration functions.

Engine Manufacturer Pilot Approach

The manufacturing company designs, manufactures and markets large diesel engines. Their approach to product development, Concurrent Product/Process Development, emphasizes the simultaneous design of product and manufacturing process. This approach requires the early involvement of suppliers in the design process and the interchange of technical data. The client has their own product modeling system, which is an Anvil derivative developed and supported internally.

The transmission media selected by the company is a public network, GE Information Services (GE IS). GE IS also provides the company with electronic data interchange capability to over 400 suppliers. The company is using GE IS' Design*Express product for the transfer of product definition data in the pilot program.

The manufacturer is redeveloping their CAD/CAM system's IGES translation software to upgrade it to support IGES Version 4. This new capability is expected in December 1989. Therefore, the pilot TDI program did not focus on testing or improving the IGES interchange between the company and their suppliers. Suppliers simply used the translated data as received.

The company implemented their gateways to the network in PC and VAX environments. The PC implementation, Design*PC, was used in the first phase of the pilot, the VAX implementation, in the second phase. The compression and data encryption was performed on all IGES files transmitted.

In the pilot's first phase, three suppliers were connected all with PC implementations and all with different CAE/CAD/CAM systems. The network was used to electronically exchange IGES files and project management schedules. The schedules were for the manufacture of subcontracted component parts used in prototype engine development.

Office Equipment Pilot Approach

The manufacturer designs, manufactures and markets copiers in a worldwide competitive environment. Under these competitive pressures, the company is transforming their product development methodology by actively involving suppliers in the development cycle. This emphasis on close ties with suppliers increases the need to send product definition data to suppliers. The manufacturer uses the Unigraphics II system running on VAX to generate and manage product definition data.

The company chose a public network, GE Information Services Design*Express, for the transmission media. They were involved in testing early versions of the Design*Express product prior to commercial release.

For data translation, the manufacturer chose to use IGES to exchange product definition data. A supplier survey was conducted to identify which CAE/CAD/CAM systems were used most frequently by their suppliers. Three suppliers, each with a different system, were selected to participate in IGES interchange testing. The three systems were:

- Computervision CADDS.
- Applicon Bravo.
- CadKey.

IGES interchange testing was performed between the manufacturer's UG II system and each of the supplier's systems. Typical design files, collected from the manufacturer, were pre-processed into IGES. The IGES files were post-processed into each target system and plotted. From problems on the plots, the intermediate IGES file could be analyzed to identify the source of the error (i.e., the pre- or post-processor). Several errors were discovered, although for the majority of the transfer tests the geometry translated accurately. For each error, the source was documented, and potential vendor and flavoring solution was described. (IGES flavoring is the processing of an IGES file to improve its reception by a particular target CAE/CAD/CAM system.) Several examples of the test results are shown below.

The following problems were found with the UG II pre-processor. These problems are significant because they cause problems in most receiving systems:

Problem #1: Use of View entity for drawing format data.

Description: UG II creates a special view just for format annotation data. Unfortunately, no provisions have been made to prevent model geometry data from being displayed in this view. The UG II IGES file indicates that model geometry is to be displayed in all views, including the format data view. See Figure 3, Problem #1 with UG II pre-processor.

Results on post-processors: This problem will cause an additional view of the model to appear in the left-hand corner of the drawing.

Flavoring solutions:

1. This is a somewhat difficult problem to address through flavoring. However, we can take advantage of the fact that each UG II View entity has its own Name Property. The flavoring software would examine the Drawing entity to see which views are active. The flavoring software would then search the IGES file for active views containing "FORMAT" in their names. Once these views are identified, we place any entities associated with these views into the Drawing entity. We then delete the Format View entities from the IGES file and eliminate their references in the Drawing entity.

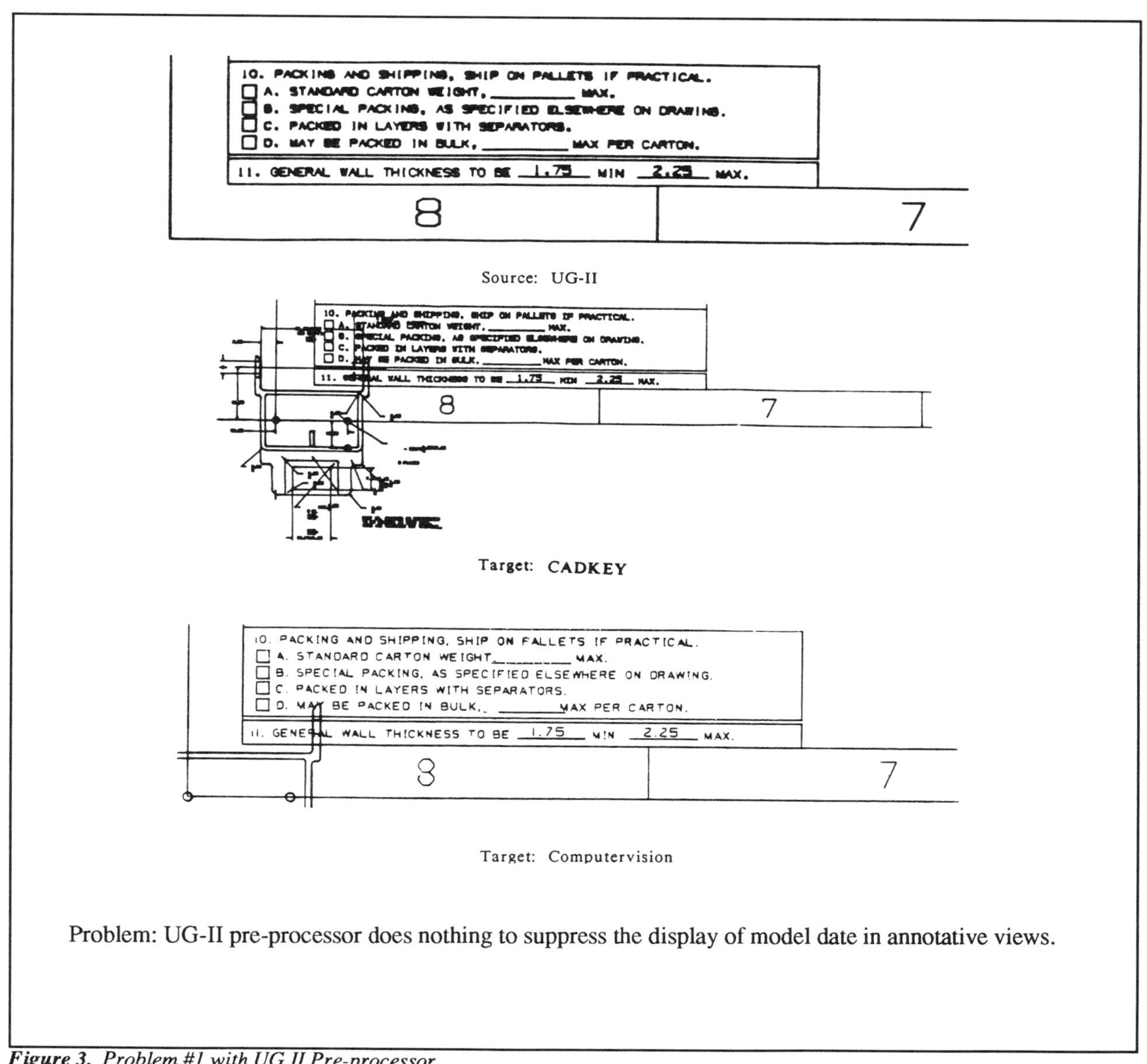

Figure 3. Problem #1 with UG II Pre-processor.

2. A simpler approach might be to utilize Views Visible Associativities to suppress the display of model data in Format Views. This would involve having the flavoring software examine the Drawing entity to determine the active views. Then the software would check to see if the word "FORMAT" is contained in each View entity's Name Property. Any views having the word "FORMAT" in their Name Property would be excluded in the views visible list for model geometry entities. In the final step, a Views Visible Associativity entity would be created containing only non-Format Views. A pointer to this Views Visible Associativity entity would be placed in directory entry #6 of any model data that currently has a blank or zero in that field. Even this solution is far from trivial.

Solution #2 is preferred over solution #1 because more post-processors are apt to support the Views Visible Associativity than they are annotation entities included in the Drawing entity.

Problem #3: Leader entities with misplaced segments.

Description: Dimension lines may have arrowheads which point in the opposite direction. Analysis of the UG II IGES file reveals that several of the Leader entities (Type 214) have misplaced segment coordinates. We define misplaced coordinates as segment coordinates that run counter to the general direction of the arrow. For example, we encountered Leaders where the first segment coordinates were to the left of the arrowhead coordinates, while the coordinates of segments #2 and #3 were to the right of the arrowhead. This is a confusing situation. UG II also created multiple segmented Leaders even when the segments were all co-linear. This is potentially confusing to a post-processor. See Figure 4, Problem #3 with UG II pre-processor.

Result on post-processors: Arrowheads are displayed backwards in the receiving CAD systems.

Flavoring solution: In Leaders, where a segment contradicts the direction of the other segments, let the majority of the segments decide the direction of the arrow. Discard the contradictory segment if possible, and decrement the cardinality of the number of segments by 1. This method would work well with the UG II files because the anomalous Leaders typically have multiple segments, with segment #1 coordinates being very close to the arrowhead, yet in the wrong direction relative to segments #2 and #3. In addition,

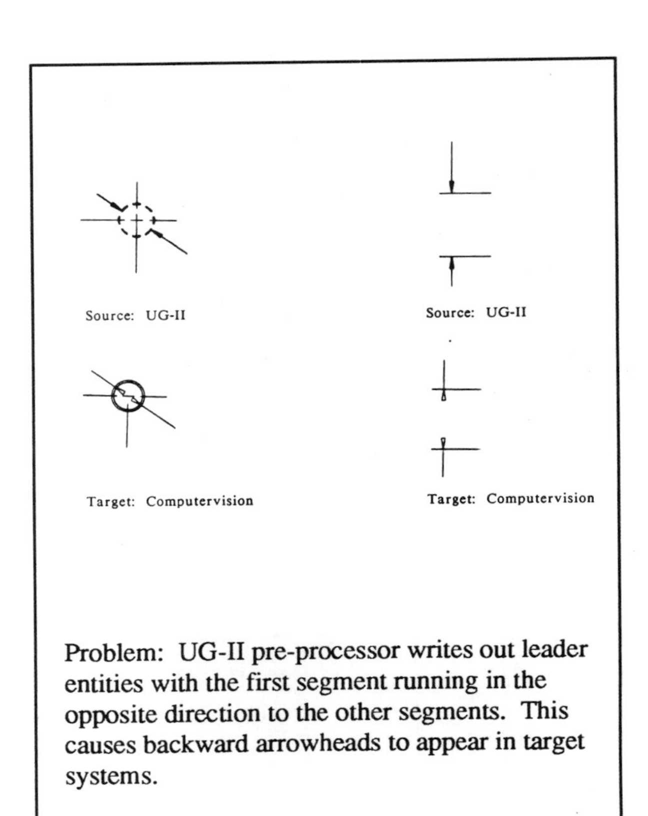

Source: UG-II
Source: UG-II
Target: Computervision
Target: Computervision
Problem: UG-II pre-processor writes out leader entities with the first segment running in the opposite direction to the other segments. This causes backward arrowheads to appear in target systems.

Figure 4. Problem #3 with UG II Pre-processor.

the flavoring software should compress the number of segments from n to 1 if all the segments are co-linear. The new leader would consist solely of the arrowhead coordinates and the coordinates of the nth segment, with the number of segments reset to 1.

The following problem was one of several found in tests between Unigraphics and Applicon:

Geometric data - The Applicon post-processor performed fairly well in translating the geometric data. In fact, the displays of the top and right views of the model look very close to the original. However, Applicon had severe problems with the transformations involved displaying the data in the front view. The Circular Arcs (Type 100's) were particularly distorted in the front view, appearing oversized and misplaced. Partial arcs seemed particularly prone to mistranslation.

Description: See Figure 5, Geometric data - UG II to Applicon.

Flavoring solution: Apply the transformation

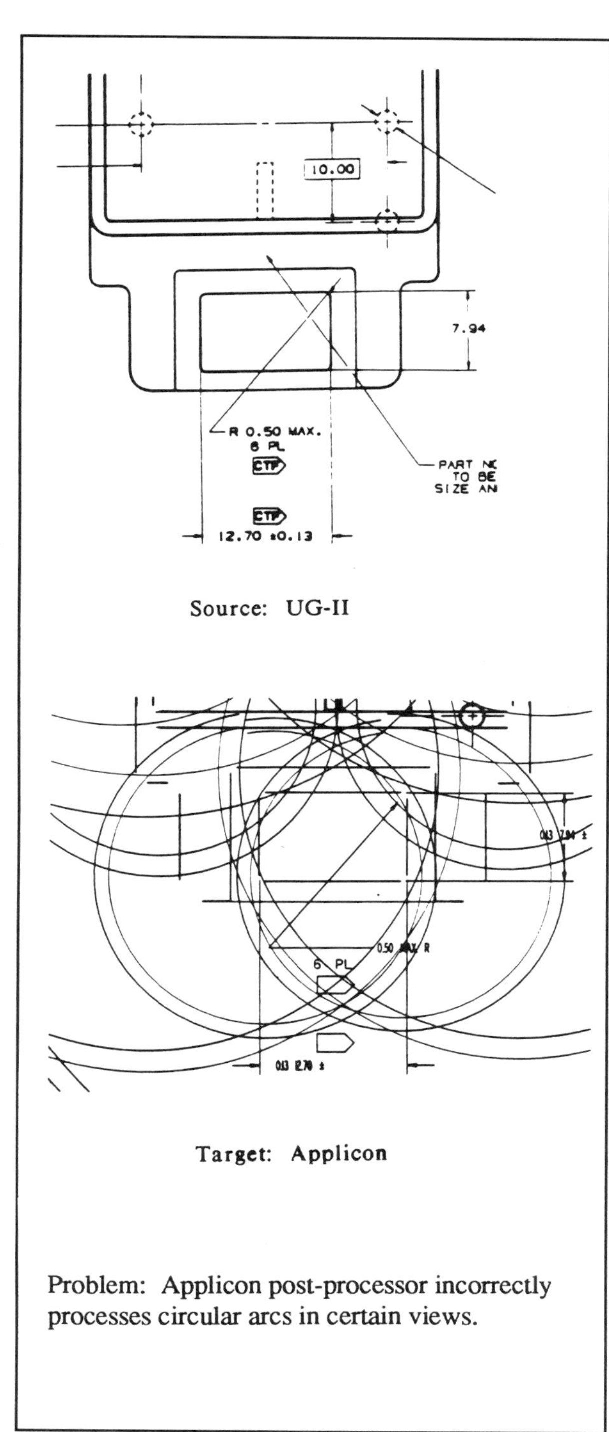

Figure 5. Geometric Data Problem with UG II to Applicon.

matrices to the IGES file, creating a flat 2-D drawing file. Let the Applicon post-processor read in the resulting file.

The last critical component of the TDI pilot program are the communication gateways. For the communication gateways to the Design*Express network, the manufacturer implemented a high speed communication service via a leased line. Product definition data was transferred from the VAX host directly to the Design*Express network. The control file containing the mailbox address for the target trading partner was generated using the Design*PC software and communicated to the network via a 1200 bpi asynchronous modem dial up connection. For each supplier, a Design*PC system was installed for the network gateway. To transfer files from the PC to their CAE/CAD/CAM systems, suppliers used alternative approaches ranging from direct connection transfer to some form of hard media transfer.

SUMMARY

Lessons Learned from Pilots

The manufacturer must be prepared to establish a pilot program that substantially approximates a production implementation of the TDI system. One objective of the pilot is to understand the challenges and benefits of production use. The pilot must have 10 to 20 suppliers connected. The suppliers must be using CAE/CAD/CAM systems which represent the types of systems found in the supplier community. The gateway to the network from the manufacturer's site must be located on an accessible, usually centralized, computer. This environment provides a pseudo-production experience for the project team allowing them to better plan for the true production implementation.

Suppliers who are asked to participate in the pilot must be contacted at a high management level. Commitment to working with the manufacturer cannot be questioned throughout the pilot. The supplier's participation is needed to establish communications, run translation tests, review programs and resolve issues. Any lack of commitment can cause serious schedule and progress delays. This lesson is especially true if the manufacturer is using the services of an external consulting organization to operate the pilot.

Benefits of Technical Data Interchange

The supplier's role in new product introduction is improved through the electronic interchange of product definition data. The supplier can take an active role in the concurrent design of product and its corresponding manufacturing process. Engineers at both the manufacturer and supplier can conduct a technical data interchange and can discuss the designs in a matter of minutes or hours as opposed to days with traditional methods. Suppliers located across the country or world can improve their working relationship by frequent exchange of product definition data. This capability allows the development team to accomplish more technical progress (iterations of design concepts) in the same or less development time thereby improving product quality.

On an economic basis, the cost of transferring the data is usually cheaper using an electronic network versus courier services. Comparing the cost of express shipping of magnetic tape to using a network, the pilot projects discovered the hidden costs of logistics associated with the tape option.

TASK	TIME
Engineer requests IGES file	10 min
Prepare to make IGES file	60 min
Process IGES file on workstation	30 min
Transfer IGES file to VAX	30 min
Find system operator	10 min
Operator copies file to tape	30 min
Deliver tape to shipping clerk	10 min
TOTAL TIME	**3.3 hours**
Engineering Time at $50/hour	$165/tape
Tape Cost	$12/tape
Shipping Cost	$25/tape
TOTAL COST PER TAPE	**$202**

In addition to this cost, the entire duration of the tasks can often last two days making the overall process inefficient and nonsupportive of the goals of concurrent product/process development.

Challenges of TDI

The accurate exchange of product definition data between CAE/CAD/CAM systems remains a challenge. CAE/CAD/CAM vendors market IGES translation software which does not fully support the entire IGES Version 4.0 specification. Their implementations contain inconsistent interpretations of vague portions of the specification which lead to interchange problems. While much effort is required to implement accurate exchange, the process must be initiated to obtain the benefits of concurrent engineering.

The integration of the various systems involved in the total process challenges level of automation achieved by the users. In most of the implementations performed in the pilot programs, the CAE/CAD/CAM systems were not connected to communication gateways. The internal systems integration achievements of the manufacturers and suppliers did not support the fully electronic handling of product definition data end to end.

ACKNOWLEDGEMENTS

The author would like to acknowledge Messrs. H.J. Ard and A. Peltzman for their contribution leading to the paper's sections on IGES Interchange Testing.

REFERENCES

1. Carringer, R.A., "Concurrent Product/Process Development: Methods and Essential Technologies,", Second International Conference on Design for Manufacturability, CAD/CIM Alert, November 1988.

2. Carringer, R.A., "Product Data Exchange Technology," Society of Manufacturing Engineers, CASA Technical Paper MS86-240, March 1986.

3. Initial Graphics Exchange Specification (IGES) Version 4.0, U.S. Department of Commerce, NBSIR 88-3813, June 1988.

4. Hauser, J.R. and Clausing, D., "The House of Quality," Harvard Business Review, May-June 1988.

5. Sullivan, L.P., "Quality Function Deployment," Quality Progress, June 1986.

Presented at the ENTERPRISE '88 Conference, June 6-9, 1988

EDI From a Supplier's Viewpoint

LEE K. FOOTE
E. I. Du Pont de Nemours and Company

Electronic Data Interchange (EDI) is rapidly replacing paper as the way of conducting intercorporate business communications. The origins of Du Pont's EDI program, the approach taken to make it successful, and its status are described. The EDI standards' issues and their long-term resolution are discussed.

INTRODUCTION

Electronic Data Interchange (EDI) rapidly is replacing paper as the means of communicating intercorporate business transactions and information. E. I. Du Pont de Nemours and Company, Inc. (Du Pont) has been among the corporate leaders in embracing and utilizing EDI to meet its customers' needs and to achieve productivity gains. The following will describe the origins of Du Pont's EDI program, the approach taken to make it successful, and its status. In addition, a perspective on the current issues connected with the EDI standards are discussed and a vision relative to their resolution is shared.

BACKGROUND

Du Pont is one of the largest industrial companies in the world with 1987 sales of $30 billion. It manufactures about 1,200 products in 200 major plants in 50 countries and markets those products worldwide. With such a breadth of operations, Du Pont is highly aware of the costs and delays associated with the traditional paper-based ways of communicating business information and transactions (such as purchase orders and invoices).

This realization first began to occur in the mid-1960s in the Transportation Department which not only was responsible for corporate logistics, but also for payment of freight bills. Even though Du Pont was considerably smaller then, this was a huge paper-intensive responsibility which they were having difficulty fulfilling within the Interstate Commerce Commission's (ICC) timely-payment regulations. In looking into their procedures, the Transportation Department staff concluded they were incurring significant costs and time delays by reentering data into their computer system from documents prepared by carriers' computers. "Why," they asked, "couldn't the information go directly from computer to computer, via telephone line, and thus avoid printing, mailing, and data reentry costs and delays?" In discussing this concept with other Transportation Managers, interest was found in examining the feasibility of using electronically communicated transportation documents. This led to the 1968 forming of the Transportation Data Coordinating Committee (TDCC) whose objective was to develop the data format standards and computer software necessary to exchange freight documents on an electronic basis (which concept was named "Electronic Data Interchange"). Over the next decade, representatives from the air, motor, ocean, and rail carriers and industry met frequently to develop standards for each of the carrier modes. In addition, a few of the participating companies provided programmers and funds to develop EDI standard transmission/reception software. All these efforts were finished by the late 1970s.

In the early 1980s, we and Chemical Leaman (our largest bulk chemical motor carrier) agreed to conduct the first pilot test of the freight invoice, and after considerable systems development effort and lengthy preliminary testing, the pilot was conducted in 1983. It was successful enough that in 1984, we went "live." That was an industry first. Now Du Pont is receiving some 2,000 electronic freight invoices, each business day, from about 40 rail and motor carriers. This has reduced the front-end freight payment processing cost by 50% and enables us to comply with payment terms for all undisputed freight invoices.

Thus, the genesis of Du Pont's EDI activity was a staff department effort—one that went unnoticed by the rest of the company.

However, while the TDCC group was focusing on

transportation-related standards, another group (which became the American National Standard Institute's Accredited Standards Committee X12) began developing the standards for business transactions (purchase orders, acknowledgements, invoices, etc.), and in 1983 published their first standards.

Something else was happening, too. American industry—especially automotive and textile—was getting battered by foreign competitors which benefitted from lower labor costs. These industries realized that to survive they had to find ways to reduce their costs. EDI, they concluded, not only would help reduce transaction processing costs (paper, mail, data entry, etc.), but also would enable fundamental purchasing changes, such as just-in-time inventory management which significantly can reduce warehouse space and working capital tied up in production materials. Those industries adopted aggressive programs to get their suppliers to use EDI. Du Pont as a major supplier, was among those first asked to participate. Thus, our focus suddenly became a marketing driven one to meet customers' needs on a high quality, timely basis.

DU PONT'S EDI PROGRAM

Responding to the needs of Du Pont's businesses, the Information Systems Department set up an EDI organization, headed by a manager (the author) with not only systems experience but also business background. This team spent the first few months determining an EDI "vision" and the strategies and tactics necessary to achieve it. This was done in consultation with all the key internal businesses as well as an outside consultant. The "vision" was that "eventually virtually all forms of customer and supplier paper communications would be replaced by electronic means, and the ability to do so would be as mandatory as the telephone is today." With that, the issue was not whether to have a comprehensive EDI program, but what its scope and pace ought to be. It was recommended that it was timely to have a broad-based, aggressive EDI program. This was approved and launched in late 1984.

A vital element of Du Pont's EDI strategy was to adopt and promote the national and international EDI standards. This was important, because of the diversity of economic sectors served, as well as the businesses' global nature. As a first step, we joined the ANSI X12 committee and the Automotive Industry Action Group. We also saw the benefit of having a Chemical Industry EDI Implementation Committee to foster EDI not only within the industry, but also (most importantly) with its suppliers and customers. Accordingly, a group of 25 chemical companies were invited to spend a day in Wilmington to discuss EDI and determine whether an industry-wide effort would be worthwhile. The group concluded it would be, and thus CIDX (Chemical Industry Data Exchange Committee) was formed, with Du Pont's EDI manager chosen to be its chairman.

Since late 1984, Du Pont has made considerable EDI progress. A comprehensive mainframe (IBM MVS-IMS) system was built to handle transmission and reception of EDI messages; validation against standards; interpretation and generation of standard messages; communications with application systems; message logging; backup and recovery, and security. At this writing, Du Pont is sending and/or receiving these EDI transactions: purchase order, purchase order acknowledgement, material release, advance ship notice, invoice, remittance/payment advice, bill of lading, freight invoice, and quality control data. In addition, customers are offered a variety of specialized information services for which standards do not exist fully (if at all); these include: electronic mail, quality data, CAD/CAE, order status, materials specifications, expert systems, and business management tools.

Thus, the "vision" of comprehensive electronic communications with trading partners is becoming reality, albeit at a pace which will always (by definition) be slower than desired. This is due to two situations: first, not many trading partners are EDI capable (it is estimated that only 5,000 U.S. businesses have EDI capabilities, yet Du Pont has over 300,000 active trading partners); and secondly, standards do not exist for all the types of information we would like to exchange.

STANDARDS—BLESSING OR BANE?

It was mentioned previously that Du Pont set a strategy to adopt and promote the national and international EDI standards. In a perfect world, that would be a perfect strategy which would minimize our EDI systems and staff costs. However, this is not a perfect world and the standards situation is a clear example of it.

Everybody accepts the need for standards—but often with the proviso, "standards are great, let's use mine." This is evidenced by the multiplexity of standards organizations which exist, e.g., ANSI X12, TDCC, UCS, WINS, NACHA, UNE/ECE/ISC(EDIFACT), etc., some of which (ANSI X12 and EDIFACT) are intended to meet everyone's needs and others of which (TDCC and UCS) are intended to meet individual industry's needs. This is a problem for companies which deal with many industries, as they find themselves in the position of having to support many incompatible standards which increase our systems and staff costs. While I expect the various U.S. EDI standards organizations to continue for some time to come, I see them eventually converting to ANSI X12 standards and changing their focus to implementation issues (versus standards development and maintenance) which will concentrate on usage conventions.

One standard issue under study now by ANSI X12 is how to migrate the X12 standards to those being promulgated by EDIFACT (the UN/ECE/ISO-sponsored EDI standards organization). The issue here is that most ANSI X12 members do not need the international standards and have no desire for those to be changed, as such would entail a large conversion effort. However, multinational companies do need the international standards, and I feel that we will spur the eventual migration of X12 standards to full EDIFACT compatibility—most likely with the X12 standards becoming a subset which does not contain the data segments and elements needed for international trade (e.g., currency and customs information).

Another major standards problem area has been the use of company-specific (proprietary) EDI standards, particularly by the major automotive and retail companies. In their defense, those were developed before the ANSI ones emerged and represent a significant investment, not only by those companies but also by their trading partners which currently use them. However, for companies just starting with EDI, to find that their EDI software (usually purchased) cannot handle the proprietary formats of major customers is distressing. This causes special systems development, customer-by-customer, which is expensive and imposes an ongoing maintenance burden. Many companies have been exposed to that situation. The silver lining, though, is that the major companies which have developed and imposed their proprietary EDI standards on their trading partners (usually suppliers who must accommodate a major customer) have realized that it's in their best interest to convert to ANSI standards, as in the long run it will lower their system and staff costs (to maintain the unique software and standards) and help keep their suppliers' costs down. Many, if not most, of these companies have begun efforts to convert to ANSI X12 standards while retaining for an appropriate period their proprietary standards capability (so as to give their suppliers adequate time to convert to X12, too). Within just a few years, one could expect that proprietary standards usage generally will have ceased.

Standard identification of trading partners is a problem, too. Some industries have adopted the DUNS number as the sender and receiver identifier, often adding a suffix to identify the internal location involved; others have gone with the Uniform Code Council's company codes; while (even worse) some use their internal vendor or customer codes which, of course, are not standard to anyone. These force many of us to develop systems to convert codes to and from those used by trading partners. This, too, is costly and difficult to maintain. I expect that gradually companies will convert away from using their internal codes to using the DUNS numbers, unless they are retail oriented, in which case the UCC codes most likely will be chosen. (An even worse situation is the lack of standard industrial product codes—unfortunately, this probably will continue indefinitely.)

Even where a company is fortunate enough to have all its EDI trading partners on the X12 standards, there are still standards issues to be addressed. As of this writing, ANSI has published official EDI standards twice (1983 and 1986) and the ANSI X12 Committee has published (1987) one update (release) to the 1986 official publication version. The X12 Committee intends to publish releases (containing all new X12-approved standards and updates to existing standards) semiannually. For X12 users, this presents a dilemma. Should they adhere only to the official ANSI standards? If so, they risk not being able to exchange transactions with companies which elect to implement the X12 Committee's releases—this is a particularly contentious issue where customers are involved. On the other hand, the X12 releases are not official ANSI standards (because these have not been subjected yet to the full public review process required by ANSI), so some trading partners may choose not to implement those—again a contentious issue where customers are involved. There is no easy solution to this dilemma, because the ANSI standard review and publication process takes at least two years (from start to finish), yet the need for prompt standards updates will continue for quite a while, until the standards have been used enough to get the practicalities reflected into the theoretical designs. This is very important, since EDI is still embryonic and the standards just now are getting to be used extensively (some, even the officially published ones, have been barely tested—if at all). Therefore, I feel X12 must issue frequent releases which have gone through the full X12 review and approval process, but which have yet to go through the official ANSI review and publication process. My suggestion to EDI users is to support the two most recent X12 releases (only one at this writing) and the most recent official ANSI publication. While this imposes some extra burden, it should enable companies to meet their own and their partners' data needs. (This is the convention the chemical industry has adopted.) The extra burden will involve having three sets of EDI tables to support, and possibly, three different interfaces for each affected application system. Most EDI software providers are committed to being able to handle at least three sets of EDI tables.

A standards question which is being considered actively now is the role of the ANSI X.400 electronic messaging standard relative to EDI. Although the ANSI X12 (communications subcommittee) has not opined yet as to whether and how X.400 will operate in the EDI arena, I strongly believe that X.400 should be the communications standard for EDI, as it also will handle internetwork exchanges and the other forms of electronic communications (mail, etc.) that companies want to do with one another. Therefore, companies would be well advised to understand X.400 and plan for its implementation as the major computer vendors introduce their X.400 software.

Last in the area of standards, there is the problem that not all needed standards exist or, if existing, are not comprehensive enough to meet today's needs. A couple of examples are material safety data sheets (being developed), compound (voice, text, and graphics) documents, and CAD. The needed standards are being or will be developed, but their completion times will tend to be considerably longer than experienced previously, because they are much more complex than the routine business transactions which already have been standardized.

SUMMARY

Standards are vital—especially to diversified companies. The fortunate thing is that most companies have come to this realization, so the resistance to national/international standards has weakened considerably. We will see the day (and not too far from now) when companies will be able to exchange business transactions worldwide on a standardized basis. However, the standards are still evolving; as users gain experience with their design flaws, they are being corrected. Rather than being frustrated by this, companies should view it positively and plan to be flexible in these early EDI years—this is a typical situation that innovators must face and is a price one must pay if interested in being a leader.

EDI is poised for rapid growth. The interest level is high; the basic standards work well; cheap and reliable software is available, and real productivity and cost savings are being achieved. Companies not currently involved in EDI should start planning their EDI implementation program now—because it will not be too long before EDI capability will be a mandatory condition (virtually now in the automotive industry) of doing business, just as having a telephone is today. A word of assurance: there are no technical barriers to EDI—it is straightforward batch computer systems technology. Therefore, EDI implementation is simply a management issue of "when."

The Success of Customer/Supplier Information Exchange

MARY K. JOHNSTON
IVAC Corporation, Subsidiary of Eli Lilly Corporation

The exchange of needed information continues to be a challenge, even in this age of worldwide communication networks. The computer age has assisted movement of data from one location to another, both internally within an organization and externally among various plant sites or other businesses worldwide. This data and information transfer is know as Enterprise Information eXchange (EIX). EIX as a whole includes a number of popular components including EDI (Electronic Data Interchange), TDI (Technical Data Interchange), Electronic Mail (E-mail), and a number of others. We begin with understanding the most recognized type, EDI.

WHAT IS EDI AND THE VALUE OF STANDARDS?

EDI, although a frequent topic of current discussion, has been available since the 1960s. EDI has been successfully used to transfer business transaction data such as orders, money transfers, shipping notices, etc. An example of a business transaction is an order received at various sites being transmitted to warehouses via a commercial network. Another example includes information related to shipments completed and returned goods received being transmitted back to the order entry location. These and other business data transactions are communicated or exchanged externally between organizations such as suppliers, customers, banks, and plants. The data or information is sent and received in a standardized format over a communications network. This exchange presupposes compatibility of the data transmitted and received. Consequently, both the sender and receiver must carefully plan applications, data formats, and data content prior to implementation.

To assure accurate data transmission among several companies, leading groups have established standards for the data format and for the exchange process. Skibinski (4) notes the importance of standards is to "establish strategic baseline expectations in regard to performance and long-term goals." The foundation to build a solid system starts here. The leading groups that have established EDI standards include the Transportation Data Coordinating Committee, American National Standards Institute (ANSI), and others such as banking, insurance and grocery industry groups. Standards eliminate the need for companies which wish data exchange to devise their own unique data exchange standard. Data and information exchange standards require less work, have a faster start-up, and result in lower costs than that in the past.

The standards are then applied to the physical communications media which currently include telephone networks (including FAX), X.25 public data networks, Frame Relay (a form of packet switching), Integrated Services Digital Network (ISDN), and Switched Multimegabit Data Service (SMDS). X.12 is the national standard for EDI set by ANSI. The European community's dominant EDI standard is EDIFACT. However, there is currently no worldwide standard which complicates inter-country data transfer systems. Some American groups such as Consolidated Freightways, the ports of Seattle and Tacoma (Washington) have adopted and are successfully using the EDIFACT system. Efforts to establish global policies and procedures are underway. Having standards enables the cost of EDI or any other EIX element to decrease. Whereas EDI predominately involves external business data and information, TDI applies to both internal and external exchange of technical information.

WHAT IS TDI?

TDI involves the transmission and exchange of product or service technical data and information including traditional Computer Aided Design (CAD) product geometry and documentation, CAM (Computer Aided Manufacturing) process geometry, numerical control machine codes, and process documentation. The transmission of CAD data sets typically use the Product Description Exchange Standard/Initial Graphical Exchange Standard (PDES/IGES) and Drawing Exchange Format (DXF). These standardized internal exchanges between engineering and manufacturing functions have significantly assisted concurrent engineering projects to increase productivity and shorten cycle time. Leading companies such as Boeing, Ford, General Motors, Chrysler, and others have used TDI to exchange data between suppliers and customers with resulting improved quality and reductions in cycle time and costs being achieved.

WHAT ARE OTHER EIX TYPES

There are many other types or forms of EIX that are less established and standardized than EDI and TDI. These other types include Electronic Mail and Facsimile (FAX) in addition to others. Eli Lilly, Digital Equipment Corp., Caterpillar Co., and many other worldwide companies, connect their personnel worldwide on an electronic mail system enabling information transfer among facilities and time zones. E-mail often transfers both business information and technical information, however in non-standardized formats. E-mail is being expanded to include high interaction suppliers and customers. These E-mail exchanges are normally accomplished on private communications networks.

CUSTOMER/SUPPLIER DATA TRANSMISSION BENEFITS

Many companies have found success in the use of EIX to enhance communications to external groups. Various industries such as automotive, health care, and retail, have found EIX has assisted them in the accomplishment of business goals, improving operations and enhanced customer services. Electronic bank fund transfers can increase cash flow and reduce float time. Mische discusses the need for common goals among partners in EIX. A good approach to setting up a healthy trading relationship generally involves:

- Establishing a high level of credibility and trust. This will help make scope and technical decisions easier because all parties will be committed to defining and furthering worthwhile business and technical goals.

- Constructing a vision of how EIX will work among the businesses and within them. This should identify the scope of the project and the ways the system will further the defined goals.

- Beginning the tactical process. Agree on technology: identify the messages to be sent, the targeted products, the terms of reference regarding these products, the specific operations to which EDI will be applied, and the time frames for it. (3)

Benefits to Suppliers

By increasing communications to customers, suppliers can reduce costs and increase sales. A Just-In-Time strategy reduces excess inventory at both customer and supplier and results in cost-effectiveness on both sides. Cressona, an aluminum manufacturer, has found that 95% of all orders were shipped on time, while reducing inventory. (3) The suppliers of Walmart, Chrysler and Caterpillar also use EDI with these customers in a JIT mode, resulting in substantial tangible benefits.

EDI has the added benefit of reducing paper documentation, as well as its delivery and storage. With the reduction of "paper pushing", personnel can be allocated to value-added activities. This reduction in overhead results in a more cost-effective workforce. Electronic ordering can automatically drive build schedules or inventory needs: restocking needs can be transmitted to the suppliers of the supplier.

Connectivity to the customer can benefit the supplier in two important ways: responsiveness to the customer and ease of future ordering. This easy customer communication link can result in a need being known immediately by the supplier. Timely response by the supplier to meet these needs can increase customer satisfaction. Since the connection is established with the customer and assists its processes, future orders are likely to increase, especially if competitors do not offer such as service.

Although the initial expense is high—an estimate of $100,000 is common—many businesses are justifying costs based on competitive advantage, cost/benefit analysis and business necessity. (2)

Benefits to the Customer

Partnering with vendors can assist customers by shortening lead times and lower costs through receiving, inspection, and quality functions, which will improve information availability and ordering capabilities.

Placing orders via EDI reduces the need to type and

mail purchase orders. This benefits the customer by saving time, supplies (including purchase order printing costs) and filing. Timely response from the suppliers is possible since inventory availability data and shipment scheduling can be part of this system. Customers do not have to depend on the mail system to assure their order is received. Even if orders can be placed by phone, many customers and suppliers reside in diverse time zones, even "half way around the world." If an internal system is designed to tie inventory on-hand with automatic ordering, additional time is saved and out-of-stock situations are avoided. Likewise, if a receiving system is interactive with the ordering system, receiving clerks can easily access orders placed.

Stocking inventory can be costly: cash flow is tied up, space constraints occur. Caterpillar has connected 950 of its suppliers, resulting in a $10 million savings in parts inventory and the reduction of 16 accounts payable clerks per Mandell. (2) EDI is used primarily for purchase orders, packing lists, shipment releases and invoices. Future plans include expansion of types of data transmitted and a change to a private network established by Caterpillar itself.

Chrysler, with 98% of its suppliers on EDI, has increased inventory turnover, lowered inventory investment, and reduced labor by more than 60%. (3) The potential for overhead reduction in space needed—inventory, filing—as well as related time spent by personnel, could be significant.

Lastly and most important, an EDI system also provides timely response to customer needs. Walmart orders goods needed in their stores Just-In-Time to replenish low inventories, which increases inventory turns and lowers costs. The U.S. government is using EDI for import/export functions; purchases from commissaries to food manufacturers; inter-government and government to commercial business transactions; and the Department of Defense (DoD) and vendor Express Payment System through the Treasury Department, which passes corporate trade payments to the banking industry.

WHAT DOES THE FUTURE HOLD?

EIX will continue to change and progress as it has over the last 30 years. Cerf (1) envisions competition between electronic mail services, fax and voice mail as well. There will also be legal issues involving the degree to which EDI transmissions are legally binding, the privacy of information and who it belongs to, and the validity of electronic transmissions and documents. International standards will be established to allow increased ease of global transmissions.

SUMMARY

The EIX technology can greatly enhance an enterprise's operations within and between companies. Current benefits include reduced costs, ease of ordering, timely response, reduced paper, reduced clerical activities, increased inventory turnover, decreased cost of inventory, increased JIT environment, and the improved connectivity of partnering organizations. As companies leverage this strategic capability, companies should pursue the following to assure business success:

- Agree on standard procedures for doing business with each other, reducing confusion and decreasing the time to process orders.
- Measure both supplier and customer satisfaction related to the real-time access to key information.
- Work to improve business relationships through EIX links; a company requiring EIX compliance among its business partners will be more likely to improve their information exchange.
- Share information with partnering companies. If a manufacturing company can better share key product availability information with its distributors, it will help everyone involved in the process, as well as the bottom line. (3)
- Share EIX benefits with partnering companies for mutual competitive advantage as tangible process improvements are achieved.

REFERENCES

1. Cerf, Vinton G. "Prospects for Electronic Data Interchange." *Telecommunications*. Jan 1991. pp. 57-60.

2. Mandell, M. "EDI Speeds Caterpillar's Global March." *Computerworld*. Vol. 25, No. 32. Aug. 12, 1991. p. 58.

3. Mische, Michael. "EDI Strategy: Businesses Shift from Technical to Business Goals." *Chief Information Officer Journal*. Vol. 5, No. 1. Summer 1992. pp. 38-41.

4. Skibinski, John. "Automated Information-Sharing Cuts Time-to-Market." *Manufacturing Systems*. Vol. 10, No. 5. May 1992. pp. 60-64.

BIBLIOGRAPHY

Cahn, David M. "What Benefits Will Global EDI Offer?" *Transportation & Distribution*. Vol. 33, No. 6. June 1992. pp. 63-64.

Frye, Colleen. "EDI Beginning to Stretch National, Business Bounds." *Software Magazine*. Vol. 12, No. 6. May 1992. pp. 88-91.

Laplante, Alice. "Troubled Automakers Rethink IS." *Computerworld*. Vol. 26, No. 3. Jan. 20, 1992. pp. 67-68.

Mason, Janet and Lisa Davidson. "Industrial Products - Caterpillar's Earth-Shaking Reorganization." *Computerworld Premier 100 Supplement*. September 1991. pp. 44-48.

Merrick, Lew. "A Better Way to Send Engineering Drawings." *Machine Design*. Vol. 65, No. 13. June 25, 1992. pp. 44, 46.

Weimer, George, Bernie Knill, Beverly Beckert and John Teresko. "Integrated Manufacturing: Systems Integrators Put it All Together." *Industry Week*. Vol. 240, No. 22. November 18, 1991. pp. IM3-IM15.

Section Six: Information Technology Enablers and Systems Planning

Section Six papers are:

- *Information System Architecture and Enablers for Enterprise Information Exchange.*
- *Integration of Product Information Residing on Various Computer Systems.*
- *Distributed Databases in a Heterogeneous Computing Environment.*
- *Which Network is the Right One?*

A description of architecture and enablers to assist information exchange and CIM environments is related by Quint. His references are basic to system technology planning for both current and future projects.

Patel maintains that product information should be accessible and transferable to various computer systems. The maintenance of this data's integrity has been a challenge. He describes the change process within an organization to enable the implementation of this sharing of information. Patel's technical discussion of a heterogeneous environment serves as another technical system architecture framework needed to respond to the systems configurations existing in most companies today.

At the heart of an EIX system is the database. Alternative software architectures and the use of distributed heterogeneous hardware platforms allow future flexibility but are difficult to implement. Rowe describes a distributed database system and its architecture that will bring about a successful heterogeneous computing technology environment.

Lastly and most importantly, Ames discusses the use of networks to ensure the exchange of vital information. He discusses what is needed prior to implementation and the various networks available. This basic document rounds out the technology areas needed to be understood for effective planning, and leaves only the need to sell management on the system's financial considerations.

Information System Architecture and Enablers for Enterprise Information Exchange

M. J. QUINT
Perceptive Solutions Incorporated

This paper summarizes the trends in information system architecture and its enabling technology as an infrastructure for EIX (Enterprise Information eXchange) and CIM (Computer Integrated Manufacturing) environments. The architecture included processor topology, network, data, applications, language, standards, and human machine interface subsets. The enablers include processor, memory, network, and interface technology advancements. This architecture and the Information Technology enablers will provide the summary basis for planning future computer based systems to exchange enterprise information.

INTRODUCTION

The Information Systems Architecture describes the organization or framework of the Information Technology components on which business information solutions are built. The current Information Technology strategy, on which the architecture is based, is to:

- Enable users to quickly and easily access and use information systems;
- Achieve improved cost effective delivery of Information Systems capability as technology changes;
- Position for the long term technology and business changes, and
- Become vendor independent.

The architecture presented here is an example of current thinking, but by no means is a perfect solution for all enterprises. It does address most issues encountered by medium and large organizations, and will assist both the novice planner and seasoned analyst when used as a guide or set of considerations.

Before an Information System architecture is created, a review and understanding of the related Information Technology enablers is needed.

INFORMATION TECHNOLOGY ENABLERS

Information Technology enablers are those elements which will allow the expanded effective use of Information Systems to occur. These are the improvements which make Information Systems less costly, easier to use, more responsive, more reliable, or more capable. Key enablers include processor chip, memory storage, communications networks, expert systems, and user interface technology along industry standards.

- Processor Chip Technology: Computer workstation processor chip speed is expected to exceed 2,000 MIPS (Million of Instructions Per Second) by the year 2000. This increase (100 times) over today's processor speed is anticipated to cost the same as today's slower chips. This power improvement will make workstations and personal computers more capable, performance-effective, and cost-efficient than most of today's centralized processing architectures.
- Memory Storage Technology: Memory technology will continue to provide higher amounts of storage in smaller devices. It is estimated that a two-inch cube will be able to store the contents of 100,000 books by the year 2000. This improvement will allow more data, information, and knowledge to be stored at the workstation or

personal computer for faster access and greater capability.

- Communications Network Technology: Network speeds today will also increase 100 to 500 times by the year 2000. Communications networks will become a transparent utility for worldwide movement of data and information in audio, video, image, and text formats. Because of the extensive use and lower costs, global networks such as EDI (Electronic Data Interchange) will expand significantly.

- User Interface Technology: User Interfaces will be significantly easier to use as the interfaces expand to include more intuitive graphics, an added "learning" function which derives and anticipates interactions based on a users past interactions, and an intelligent module which adjusts the interface based on user behavioral input.

- Expert Systems Technology: Advances in computer languages and procedures for capturing, storing, and presenting expert knowledge will continue to increase. These large programs will become very practical as memory and processor chip improvements occur. Expert systems technology will foster commonplace "intelligent or smart" applications, user interfaces, and machines by the turn of the century to support user decision making and cognitive processing.

- Standards: The ability to leverage a wider scope of technology advances is essential to the business strategy of the 21st Century. Standards provide the way to take advantage of other technology components. Standards will continue to be both demanded by industry consumers and accepted by technology suppliers. Standards will accelerate technology advancement as a whole.

These Information Technology enablers will have a substantial impact on the architecture designs positioned for future systems. Having an idea of the improvements anticipated, the Information Systems architecture now can be addressed.

INFORMATION SYSTEM ARCHITECTURE

An Information System architecture includes processor, network, data, language, application, and human-machine interface architecture components. Each component architecture may be described separately, but must be integrated into the complete Information System.

- Processor Architecture: The increased computer processor power and increased memory available in a smaller size at lower costs has made the client-server processor topology price/performance effective. Client-server architecture will replace the centralized processor architecture over the next five years, especially for applications supporting interactive and compute intensive users. In this architecture, a client (workstation or personal computer) cooperates with a server (large processor or mainframe) to perform a task. The client (requestor) performs the user interface processing and some or all of the application processing. The server (supplier) provides the remote data, some or no application processing, and application and network integrity and security. The client-server operating system of choice for future positioning is Unix and it derivatives.

Workstations will use RISC (Reduced Instruction Set Computer) technology. RISC technology has eliminated some excessive processing overhead and resulted in faster and less expensive workstation applications.

Lastly, fault-tolerant or fail-safe server processor technology is becoming cost-effective and will be commonplace by the year 2000. Fault-tolerant processing will give the appearance of a system that never fails, which should increase the system user's satisfaction and productivity.

The above client-server configuration will result in lower net costs and increased applications performance.

- Network Architecture: The network architecture emphasizes connectivity to heterogeneous computing elements. The network architecture of choice uses Ethernet Local Area Network (LAN) and the TCP/IP protocol for connecting dissimilar clients to servers and networks to networks. Wide Area Networks (WANs), ISDN (Integrated Services Digital Network), a two-way cable service, and EDI (Electronic Document Interchange) will be common and integral components of a global network used to connect distant plants, partners, suppliers, and customers electronically.

- Data Architecture: The enterprise data and

information needs will be modeled and resultant data architecture engineered. The components of the data architecture will use Relational and Object-oriented Distributed Databases. Data dictionary and data views will enable both large transactional processing and user ad hoc queries and updates. Data integrity will be maintained by the server or host processor. Databases will contain multimedia including text, graphics, video, image, and audio forms.

- Language Architecture: Application programming languages will use standards based Object-Oriented Programming technology. Applications will largely be accomplished in the C++ language. C++ is the current preferred language because of its portability and reusability across many platforms allowing platform independence. Applications will be individually developed for the near term and automatically programmed using "lower CASE" tools in the long term. Programming languages will be used to generate reusable and multiple-call applications.

User or client programming will use either Structured Query Language (SQL) or natural languages which translate to SQL to access, manipulate, and report data and information.

Application Architecture: Applications will be engineered to be integrated with each other and the information technology infrastructure. Application subfunctions will be CASE designed to be callable from any other application or query allowing the user to effectively create application functions as needed. Expert system subfunctions will be a common part of applications. Applications will access and use a common enterprise multiview data/information/knowledge base. Application designs will be separate from data management, and user interface functions. Client organizations will tend to buy robust integrated application modules rather than develop them internally.

Human-Machine Interface Architecture: The human-machine interface architecture will be sociotechnically engineered. The human-machine interface's effectiveness will continue to expand because of having additional Graphical User Interface (GUI) functionality, a common menu subsystem (available from any platform), and "learning" capabilities customized to the individual user. The GUI of choice is Open System Foundation's (OSF's) Motif standard and Microsoft's Windows/NT product.

These Information Technology architecture elements, when aligned and integrated, will provide a flexible, efficient, and effective base for information and knowledge exchanges.

SUMMARY

In this document, an EIX architecture and its enablers have been summarized. The Information Technology enablers provide an understanding of how technology will change in the near future. The architecture, considering the enablers, will become the framework on which the EIX systems will operate successfully.

BIBLIOGRAPHY

The following references are provided as additional sources of architecture related information.

Scientific American. "Communications, Computers and Networks." *Scientific American*: September 1991. This issue is devoted to a series of articles on how computers and telecommunications are changing the way we live and work.

Scott Morton, Michael S. *The Corporation of the 1990s: Information Technology and Organizational Transformation.* New York: Oxford University Press, 1991.

Zachman, John A. and John F. Sowa, "Extending and Formalizing the Framework for Information Systems Architecture." *IBM Systems Journal*, Vol. 31, No. 3, 1992.

Presented at the CASA/SME AUTOFACT '90 Conference, November 12-15, 1990

Integration of Product Information Residing on Various Computer Systems

ATUL C. PATEL
SYSTECH, Incorporated

Traditionally, engineers and designers created product definition information (PDI) on drawings. During the 1960s and 1970s, computers were heavily used in processing alphanumeric product structure and process information. Later, in the 1980s, they kept geometry information of the product. This led to the computer systems maintaining PDI on different databases (graphics and alphanumeric) on heterogeneous computers and operating systems. This environment caused inconsistent PDI and delayed access for various departments. The product definition information system (PDIS) must be redesigned so that it can take advantage of advances in computer and communication technologies to provide more valuable information, not just data. PDI within the databases should be accessible and transferable to various computer systems while maintaining their integrity.

INTRODUCTION

The most important information any Engineering and Manufacturing company needs is the Product Definition Information (PDI) in a usable form and easily accessible. The computer industry has neglected this most important aspect of any manufacturing business. They have made computers and developed proprietary software to compete in the market with similar solutions to the end users, and at the most, have improved individual productivity of office staff and engineers. This may increase individual productivity, but unless the information, once entered into the computer, is used by others, there may not be an increase in the overall productivity. Every system will have data at the different work centers reentered. Sometimes, data, if automatically transferable, is not in useful form when it is sent electronically to another work center and may require altering data format before it can be useful to that work center. This method of operation takes up a good amount of time; integration and communication of information to various groups in any enterprise then becomes a painful experience. Unless PDI management is simplified, companies will be spending an excessive amount of resources without gaining any productivity and will, eventually, result in a decrease in revenue.

It is imperative to address the sharing of PDI among Engineers, Designers, Researchers, Manufacturers, Marketers, Purchasers, and Managers. Let us examine the way information is created and shared by various groups of an Engineering and Manufacturing company (Figure 1).

Based on customer request or market research, a product is conceptualized to satisfy desired functions. Next, a design layout is made of the product and some amount of design detailed and analyzed. It is then released for drafting (detailed design). Meantime, most of the analysis on the product has been completed and approved. In this process of product development, there is a need for various groups in engineering and R&D to correlate the product data.

Once signed and approved by the respective groups in Engineering, the design is released to manufacturing and procurement for action. Many groups outside Engineering may have started acting on the data even before the formal release from Engineering. Now the serialized process defined never existed and, though preferred, was never formalized. Later, manufactur-

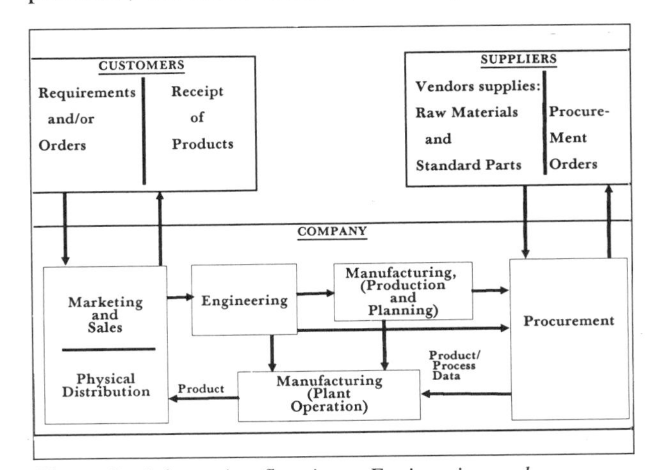

Figure 1. Information flow in an Engineering and Manufacturing Enterprise.

ing, on getting the product data, goes its own way in manufacturing the product. Sometimes it is difficult at that stage to see that the as-built product is not the as-designed product. Not maintaining the configuration control of the product will haunt any company when the product fails. They are bombarded by customer complaints, investigated for product-related accidents, and resented by the in-house engineers who have no way to make improvements on the product without the history of the failed product.

In most of the companies, barriers exist among various departments, and thus, information is not readily exchanged. Computer Based Systems bring integration, information exchange, and simultaneous design, based on the way the company sets up design and manufacturing process. These systems can provide better operations and a smoother flow of PDI in a company.

The concept of one logical Bill of Materials (BOM) should be a primary objective of any company. Changes to the design or manufacturing process should be centralized or, at least, well managed through automation and correlation so that proper change control is maintained. It is possible for the computer system to keep track of such changes. Recent advances, in very high speed processing, data access and data transfer and in storage capacity of computers, networks, and operating and database software, will provide enough support in implementing an integrated environment for Engineering and Manufacturing companies, regardless of the company size. More noticeable changes like Graphics User Interfaces (GUI), and Distributed Databases, Files, and Processors brought a revolutionary vision of developing a customized Product Definition Information System (PDIS) for every company.

With this overview, the following subjects will be discussed in order to provide some of the answers you are seeking to manage PDI on heterogeneous computers, and develop confidence in building integrated PDIS.

- Engineering Product Definition Information
  - –Enterprise information flow
  - –Information flow to a machining center.
- Computer and Network Systems
  - –Hardware and Software Technologies
  - –Off-the-Shelf-Application and Custom Software.
- Heterogeneous Computer Systems
  - –Design of the Product Definition Information System.
  - –Advantages and Disadvantages of Heterogeneous Computer Systems.
- Implementation Case Study
  - –Electro-Mechanical Industry
  - –Aerospace Industry.

At the conclusion, I emphasize that the time to start designing and implementing PDIS is, at present, using heterogeneous computers and networks.

The nature of this subject gives rise to the use of abbreviations. Appendix ''A'' lists abbreviations used in this paper.

ENGINEERING PRODUCT DEFINITION INFORMATION

Enterprise Information Flow

Figure 1 shows a schematic of information flow among groups and departments in an Engineering and Manufacturing company. Based on customer needs—recognized by market research or requested from the customer—a product is conceptualized, designed, analyzed, manufactured, tested, and delivered to the customer. Major cost centers are manufacturing and material procurement for producing the product, when compared to engineering and design. However, engineering and design directly affect the cost of procurement and manufacturing. With growing competition, it is important to reduce the products' order-to-deliver cycles in an Engineering and Manufacturing company. These products' order-to-deliver cycles, though directly controllable by the management and least affected by the outside market forces, are still not competitive enough. This has forced many U.S. companies into automation—not only for manufacturing, but also for managing PDI.

We have yet to learn how to manage the automation process for achieving productivity and reducing the products' order-to-deliver cycles. Today's computers, networks, workstations, and peripherals are capable of producing wonderful results, but integrating them for a company's operation is a challenge in itself. With available options, it is more confusing than ever and requires well-qualified professionals who can select the right computer-based system components. Any compromise in the selection of people to automate may become the starting point of Doomsday for the whole company. The bottom line of any company is its products, and any delay or mess created by automation will impact the overall effectiveness of the corporation to deliver goods to its customers. So far, experts have talked about Just-In-Time for materials to reach work centers. Now, it is equally important for the information to reach Just-In-Time at work centers. With the low cost and flexible systems in the market, it is possible to transfer information electronically to any work center where information is required.

Distributed databases, distributed processing, Local Area Networks (LANs) with network software, like UNIX, and user friendly GUI are contributing to accomplish the goals of automation for numerous companies who were not in a position to exploit computer technology until the low cost, economical, and powerful computer systems started flooding the market within the last two years.

With all this going on, let me caution you that unless you have taken time to understand your operation and need of information, many of you will still be implementing islands of automation, not only separated by an engineering and manufacturing wall, but also, within engineering, among design, development, and research groups.

Arrows among various functional groups shown in Figure 1 can be network cables or human expediters running with hard copy change orders throughout the organization. No matter what kind of mechanism is employed, the schematic reflects the flow of information in an Engineering and Manufacturing enterprise. This is the PDI flow proposed for automation, which can satisfy the need of manufacturing companies.

It is common to notice that many of the computer application programs, even from the same vendor, have separate databases, and those who are integrating need the transfer of information from one application to another by manually keying in after reading and noting them on paper. Why? Some vendors have a clever argument. They say much of the analysis on products do not require automating transfer of information from one system to another. Time is saved by directly keying-in geometry and completing the assignment rather than waiting for the time-consuming automatic transfer of data from 3-D models. For example, a structural analyst for 3D model of a part in need of analysis will prefer to geometrically produce the structural model from scratch rather than use the 3D geometry from the 3D modeling program. This partly may be due to lack of functionality of applications and systems software, and it has yet to develop the capability to create a true structural model from the 3D design model. Also, required computer resources which can perform analysis may be very high. In such a scenario, though sometimes the job gets done, inconsistencies on numerous occasions in the analysis, design, and manufacturing, infiltrate and cause more problems and delays than one can imagine.

There are numerous straightforward methods like IDEF0 (ICAM-Integrated Computer Aided Manufacturing, under an Air Force Manufacturing Technology project), HIPO (Hierarchical Input Processing and Output), and Entity-Relationship 1 to define the PDI flow in an enterprise. It is easy and desired that the PDI flow is determined before embarking on the automation and implementation of PDIS on computers. Wiederhold and Elmasri[1] (1980) and some of the papers stated in the bibliography are excellent sources for information on methodology of defining PDI.

Information Flow to Machining Center

Figure 2 shows the flow of PDI from Engineering to Manufacturing through various computer systems or manually to work centers. This is the present day scenario evolved through advances in diverse technologies but in an uncoordinated way to manage the PDI flow. You may have witnessed that to keep the old system and enhance it with the new computer and communication technologies, inefficient systems, as a result, have been created to manage the PDI. These monstrous systems have crippled some major defense contractors so that they cannot effectively compete with other companies that have redesigned their systems using new technologies.

Figure 2 shows two distinct paths for the information to reach a machining center. It is imperative that the PDI reaches the machining center from both paths on time. Absence of one or the other will cause the delay in executing the task at the work center. This will definitely affect the overall production schedule. Just as it is important to have materials to reach the work centers on time, it is equally important for the information to reach on time. No doubt that the engineering release function defined in Figure 2 is performed in almost all companies, but it is time-consuming and sometimes ineffective. To maintain the schedule of a department, designers have compromised the quality; inaccuracy of a design has been allowed to pass on to manufacturing under the pretext of meeting design schedule. One should realize that processing a "change order" is very expensive and more frustrating than processing the "engineering release." An efficient computerized checking system which can perform logical checks on the drawings *before* release and also at the time of design, can be a boon to the manufacturing industry. Any function performed by any group or department should consider overall impact on the eventual product development, rather than only on the task at hand.

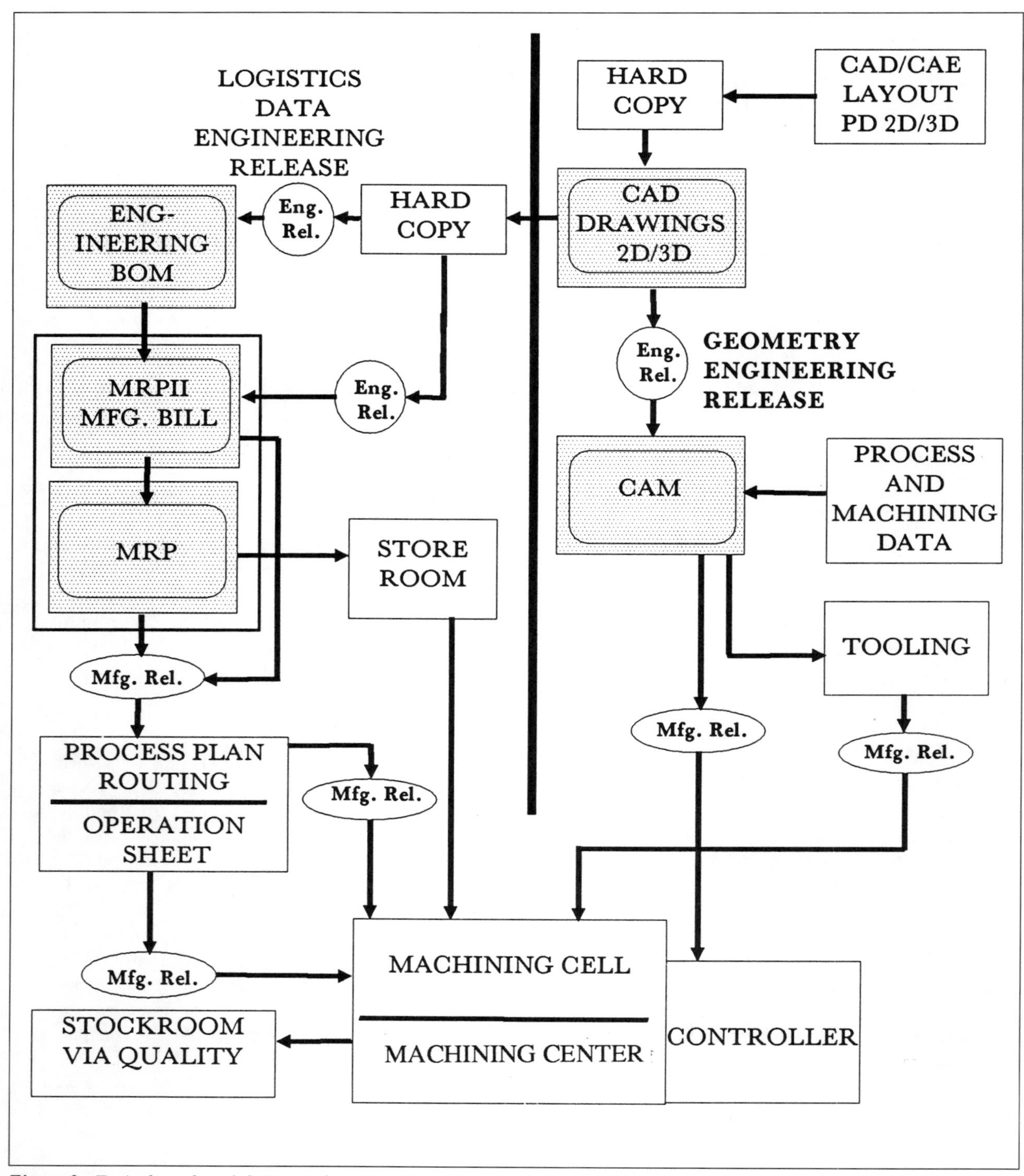

Figure 2. Typical product definition information flow to a manufacturing cell.

COMPUTER AND NETWORK SYSTEMS

Hardware and Software Technology

In the last few years, we have seen major advancements in computer technology and its uses. Numerous concepts and ideas have been put forward, and some have been implemented to create computer products like personal computers, workstations, add-on boards for additional functions, Local Area Networks, Multitasking, Multi-user Client-Server computer network systems under UNIX and MS-DOS, and communication software and hardware to connect dissimilar hardware and operating systems such as VMS and MVS/XA from DEC and IBM mainframe manufacturers. This has helped in providing engineers with workstations on their desks. Numerous database management systems have now distributed processing, database, and file environments which allow every company to design their own PDIS. Data can be stored on the database file server closer to the engineer's work center, or wherever it is required.

With advances in network hardware and software, such as fiber optics and standardized UNIX, respectively, many of the concepts which have been theoretically developed over the years are being implemented. Wiring the factory is becoming trivial to transfer data and distribute various functions to different nodes of the network. Each node may represent a work center or data storage device and any other device to communicate to the outside world through modem or produce hard copy through printers and plotters. A node on a network can be any one of the following:

- Workstation,
- Personal computer,
- Server–database or file or both,
- Terminal,
- Data collection device,
- Data storage system,
- Central processing unit,
- Peripherals–plotters and printers,
- Various databases–graphics and alphanumeric, and
- Application to application data links.

The next generation of computer systems to manage PDI will be in paperless and wireless factories where information is communicated electronically and through Radio Frequencies (RF) and displayed on the terminals and workstations in real time. This new generation of hardware and software will be flexible. In many instances, it will become disposable for continual upgrade of PDIS—unlike the predecessors where you are stuck with the initial design of the system which took years to implement.

The technology identified is commercially available at a very reasonable cost and is used in numerous applications for data collection and information processing. Now this can be used for the management of PDI.

With the increase in options of having various hardware and software products in the market, it is very important to design and implement networks to meet the customer's requirements. It is important to avoid selecting incompatible products. It is possible to create chaos in a network with incompatible products for PDI management. For instance, it is found that the Geometry data representation in one computer is not the same on another. Non-IBM and IBM file formats are not 100% compatible, even with the conversion programs converting in both directions, which can also give rise to erroneous results when dealing with geometric entities like solids and surfaces. Thus, it is not advisable to have computational geometry or other numeric data exchanged between computers using non-IBM and IBM data formats respectively. Another example of incompatibility is in UNIX operating systems: UNIX on one workstation is not 100% compatible with the UNIX on another. Using yellow pages (under UNIX) to manage network and computer users on the networks becomes difficult when they do not share files due to incompatibility at the binary representation level. This calls for experienced and knowledgeable experts to design, implement, and manage emerging network based computer systems for PDIS. Numerous workstations with powerful graphics and processing speed offer engineers to produce designs in 3D and perform analysis. A judicial choice has to be made to see that the PDI is shared by various departments and groups without re-entry of information or getting into elaborate conversion of files every time the design data is accessed.

Another technological area to be looked into is image processing and conversion of images to vector drawings and text. There are programs available which can convert simple text and graphics elements to vectors and be used to enter into the vectorized geometry database. This image-to-vector conversion and data exchange between open windows can be effectively used to extract information from the standard component libraries supplied by the industry associations, vendors, or your own company on a CD-ROM in an image format. Even the vendor files maintained in CD-ROM can be used to identify sources of parts required by the designer.

With Multimedia (Text, Database, Graphics—Vector and Raster), Multi-tasking, and Multi-user workstations, engineers will be able to access reference material on-line, accessible via company's standard libraries or industry associations' database. At the same time, information for Graphics, Database, Alphanumeric editors, and reference image data will be displayed on the screen. From here information may be exchanged between open windows on the screen.

Off-The-Shelf Application and Customized Software

Coming to application software, like release and change control, MRP, CAPP, MRPII; Geometry (3D and 2D) and engineering analysis are some of the valuable programs which can design and manage PDI. Still, they are not integrated into one, nor properly interfaced. Even CAD (3D and 2D) from various vendors are not compatible and cause difficulty in sharing information automatically on computers with other departments. Though difficult, with a prudent choice, various products compatible with each other can be selected by experts in industry for PDIS.

HETEROGENEOUS COMPUTER SYSTEMS

Designing the Product Definition Information System

Connecting various devices from different manufacturers operating under compatible software, networks, and protocol data converters like transceivers, bridges, routers, and repeaters, helps systems integrators put together excellent systems to manage PDIS. This technology is a boon to the manufacturing industries. Many companies use this technology to integrate their documentation and proposal preparation to reduce the time and effort required. From this, we can learn a lot about the technology and make every attempt to design PDIS. Automating proposal preparation is great, but *what* is proposed is equally important, especially for the Engineering and Manufacturing companies in need of improving PDI flow through the company and, if possible, to their vendors and customers.

At the international and national levels, attempts are being made to standardize the interfaces and operating systems including networks. One of the noteworthy efforts is in Europe by Esprit (a consortium of 21 European companies, including IBM) on CIM-OSI (Computer Integrated Manufacturing—Open Systems Integration). Although CIM-OSI appears to be similar to MAP/TOP—which is based on ISO standards—they have different aims. MAP/TOP is trying to get interconnection between systems in a standardized way, while CIM-OSI concentrates on communications within systems. Data should be directly accessible by applications. CIM-OSI will be built on MAP/TOP and then go beyond. This standardization project hopes to define an ''integrating infrastructure'' on which PDIS applications can be built. Methodology for modeling a manufacturing enterprise exists and it must be used to develop a PDIS model. Attempts made by ISO may result in a comprehensive standard against which products may be developed. Because of support from companies like IBM for the standard, ISO hopes to have a document which vendors can use to develop products by the end of 1991.[2]

Use of PCs and workstations with LAN has revolutionized the whole manufacturing industry. At this time, it is imperative to think and talk about the integration and intercommunication of the PDI throughout the organization. The PDI need not be available to a selected few and on paper only. It is possible, with a careful design of the network-based PDIS, to provide most valuable data residing on various databases (CAD, CAE, CAM, and CIM; Figure 3) to work centers where necessary, no matter in which database the data is stored. It is high time vendors look at the needs of the manufacturing companies in terms of PDI rather than just converting files (various computer file formats) from one application system to another application system in totality without the logical access to the PDI residing in the database (file).

A well conceived design to manage PDI throughout the product's life cycle with presently available Client-Server Architecture using workstations is provided. Figure 4 is a schematic of a PDIS scenario which can satisfy the needs of the Engineering and Manufacturing companies.

Every department will have its own database and file servers, including its application programs executed on the stored data. These databases and files will be accessible by other departments on a need basis and predefined by the management. Control of these databases will be with the respective departments who own the data. Inter-database transfer of information is controlled by owners of the respective databases and programmed into the interface programs. A well defined release system will help in implementing and executing the PDIS.

It has become clear that it is possible to design databases as well as a control (Release and Change) system, but unfortunately, implementation of this design on computer systems is difficult. Even with the compatible hardware and software, limitations of capacity and speed of data transfer of the computer systems are still major roadblocks for a successful implementation of PDIS. Exploitation of the technology, used by online real-time databases, high speed networks, and supercomputers, will be required in order to implement an effective PDIS. Though these technologies exist, they have not made inroads into the development of PDIS. It can very well be that such a system is not visualized by many engineers and managers in charge of operation, nor articulated to the developers of such systems.

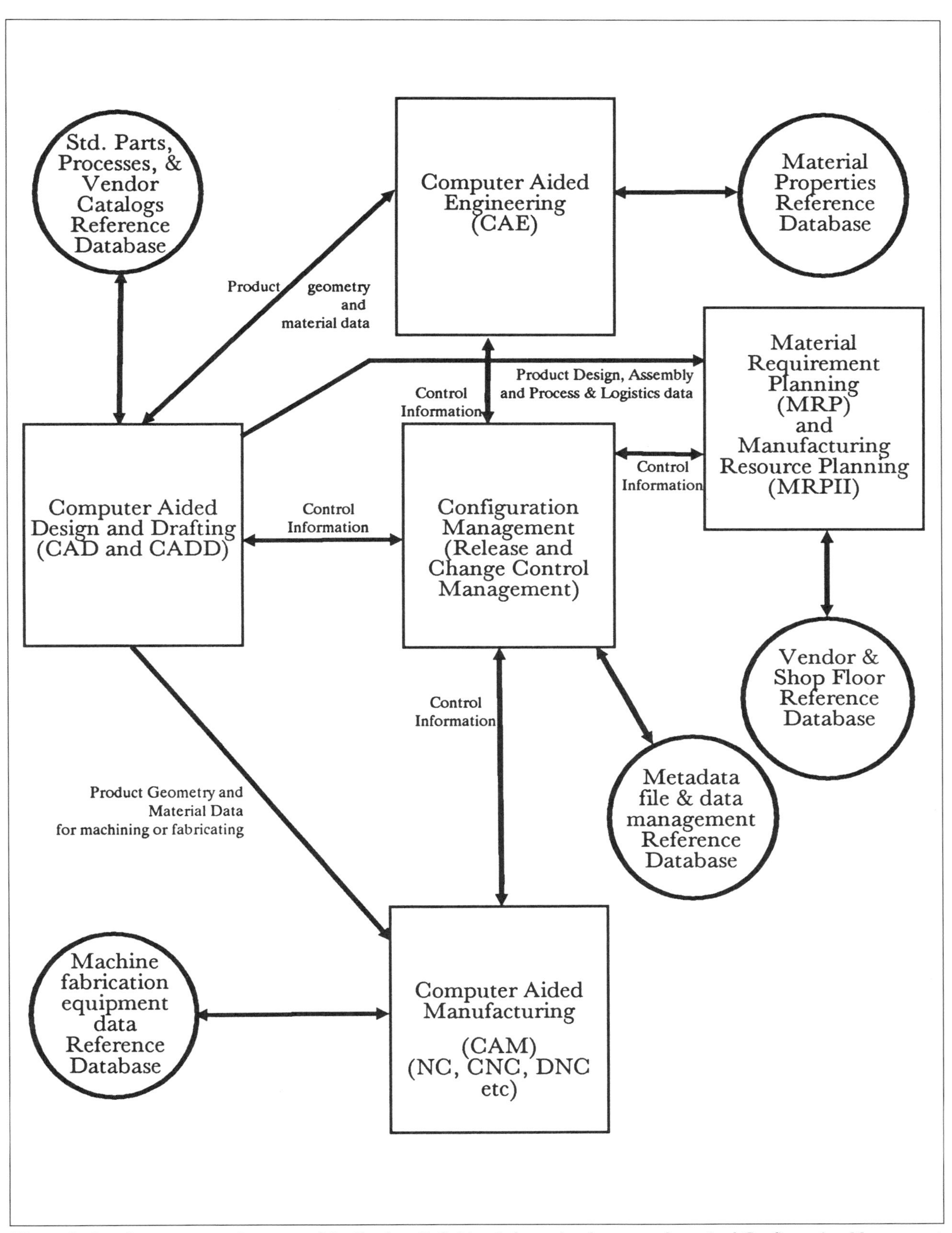

Figure 3. *Interfaces among subsystems of the Product Definition Information System and required Configuration Management (release and change control).*

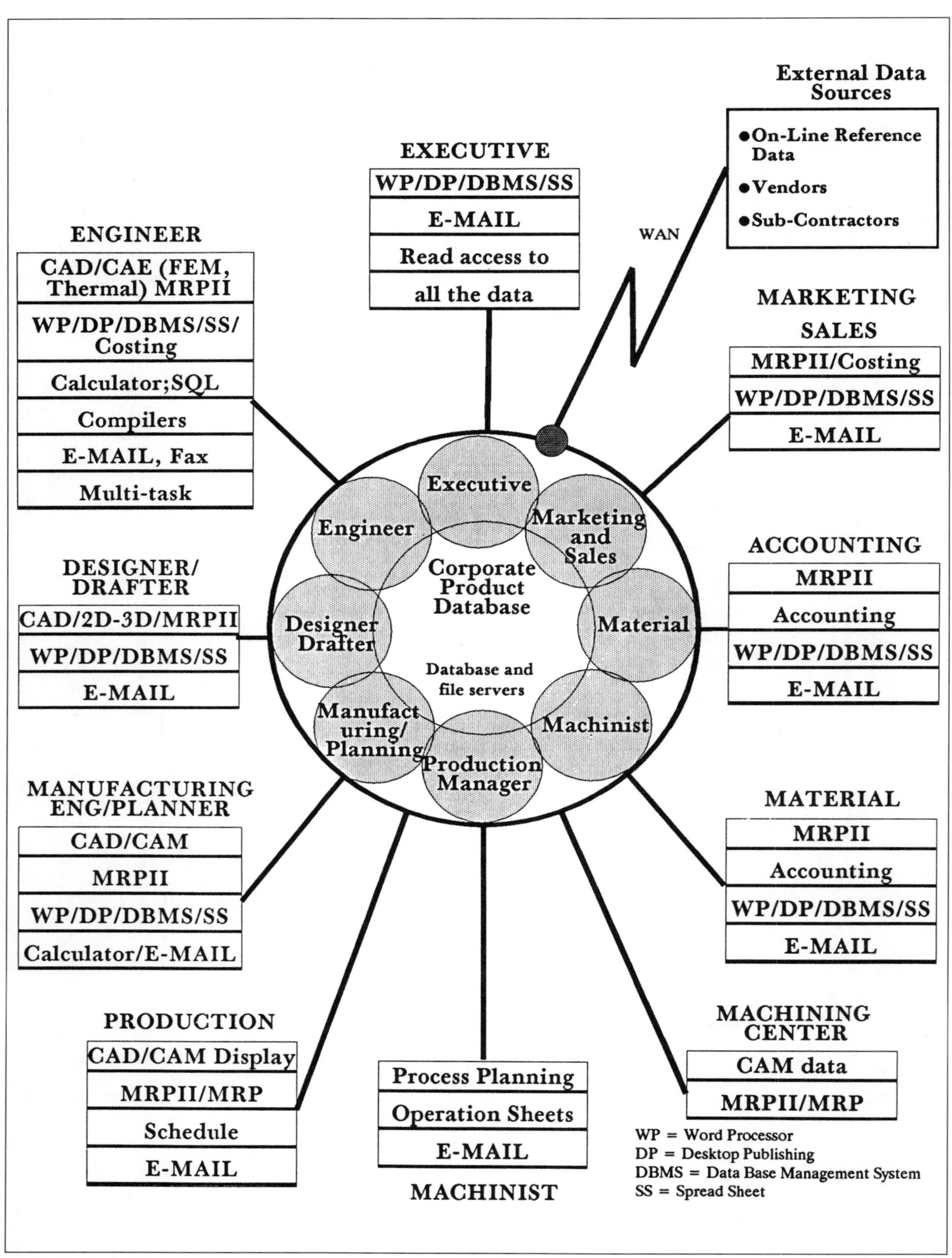

Figure 4. *Product definition system using workstations in a client-server network architecture.*

Figure 3 shows the relationship among various systems having PDI. These systems have their own physical database, but the logical data within them are not related across databases. Interfaces are required to complete the information, as well as maintain integrity. Configuration control to provide security and integrity of PDI is mandatory when the customer is the U.S. government or any other government. Even for nongovernment customers, it is crucial that the most effective computerized configuration control system be used to produce quality products. Implementation of the configuration control tracking system, which includes a release and change control function—a trivial design—is not easy. A majority of the rules and procedures are programmable and, with present computer technology, manageable. Of course, at present, there are not any products in the market to provide a shell to be built on. With the distributed databases and servers, in coordination with the operating systems, centralized information about the status of the drawing can be maintained on the database server. Database administration functions can be simplified to manage released drawings, changes, and Was/Is (change) information.

A system of storage, including archives, can be designed to provide the up-to-date status of the PDI in a system. A repository in the mainframe or customized system or database server on workstation-based networks provides an excellent tool in managing PDI. It is important to point out that a method of managing releases and changes on drawings and parts lists is crucial. With an efficient method of managing changes, the Change Orders may not be required; changes can be managed by Revisions. This will be possible because of automation, and it is least troublesome to produce new revised drawings and parts lists with Was/Is reports, automatically. DoD requirement of making new revised drawings after few (suggested around five) change orders, though mandated, is not practiced and some of the companies end up with unreasonably high numbers of change orders before they make a revision. To expedite and reduce work, many design managers have embarked on making master drawings with a number of parts lists representing different assemblies. This practice was good when the system was manual; although it is not wise to automate the manual method, ignore the power of automation and an access to the up-to-date information on PDIS.

There are numerous commercial products available which can help in developing PDIS. Many vendors of RDBMS and OODBMS provide good infrastructure to develop release and change control systems which can help in activating numerous processes which are supposed to take place at the time of release. On release, part information is transferred to Manufacturing and Materials; logistics data are fed automatically into MRP and MRPII systems; geometry data is fed to CAM systems, and the status of the PDI is maintained without physically moving data files from one database to another. At the time of release, a Was/Is report can be produced and stored in a database. Status of the data sent to CAE and CAM systems from CAD can be recorded on the centralized control database, and finally, effective changes can be broadcasted to all the concerned departments through the use of Electronic Mail (E-Mail).

Furthermore, Interfaces among various systems can be programmed to transfer PDI to various systems conforming to the transferred systems' database formats on a controlled basis. There are numerous questions raised in terms of timing of the data available to manufacturing about any variations manufacturing would like to make. To control the product configuration, the data should be well controlled, even though changes can still be efficiently performed. With all the streamlining, the systems designer must accept the fact that changes to the PDI, as well as to the system, are inevitable. In numerous cases, the computer system can lend a hand to the present manual system, especially when the design of a product is changed due to a defective component. Based on the progress of the product design and manufacture, different procedures are invoked. They are managed by the change control board and involve rework, retrofit, stop work, inactivation, and restriction. The existing manual system is inadequate to handle such changes since manufacturing planners have to go through all the drawings to locate affected parts. With the present day technology, this job of looking through all the drawings and assemblies, and preparing a list of components, assemblies, and drawings that require attention, can be prepared easily with all the details of where each part is in the shop, which can then be tracked by a subsystem of PDIS.

The history of all changes (Was/Is report) can also be managed by storing it on-line and logically placing it in archives on tapes or WORM disks, or CD-ROMs.

With the present day technology, it is even possible to produce process plans and work orders automatically. Even change orders can be printed with the sketches on the same document for clarity by document programs which can be a part of PDIS.

The design process of PDIS is a cooperative effort of a whole company. Each department plays a role in defining its needs. A Systems Integrator function has to be created in order to select proper compatible products to implement the specifications. Interface design, configuration management, data translators, and displays will have to be coordinated by Systems Integrators and prudent choice has to be made. Though operations management are the ultimate users, they should not be burdened with selecting technically compatible products and interfacing protocols. Their opinions should be valued as long as they do not unreasonably violate the objective of systems integration and become an obstacle in the development of PDIS .

Advantages and Disadvantages of Heterogeneous Environment

Advantage

Possibility of mixing hardware and software from various vendors which are compatible for cost effectiveness and functionality.

Comments

Availability or development of interfaces, file format standards, and data conversion programs give this opportunity. Though not perfect, this capability gives rise to excellent opportunities to design PDIS. At present, users are at the mercy of computer vendors, and thus, use capabilities which are not adequate for an effective PDIS in a heterogeneous environment, provided by these vendors. A prudent choice of integration products can help achieve improved overall results in developing PDIS.

Advantage

Utilization of the capability of products offered by various vendors specializing in an area of PDIS because no one company may be able to provide the best products in all the areas.

Comments

Companies strong in Graphics may not have products which can satisfy the needs of engineers in analysis, industrial engineering, process planning, or numerical control. In such cases, using a multi-vendor system may be the answer.

Advantage

Maintain databases locally and allow access to other areas on need basis while maintaining control.

Comments

Distributed file and database management systems are available on the Client-Server Architecture, and servers can be dedicated for a particular area of the PDIS data. Powerful SQL also plays an important part in developing and maintaining PDIS.

Disadvantage

Present networks have slower data transfer rate.

Comments

All of the computers for PDIS will be linked with other machines, but the networks aren't getting any faster. As a result, the raw speed of the new workstations is wasted when PDIS share or swap data—which occurs often. Current standards for networks, such as Ethernet and Token Ring, may not be adequate. The current desktop computers used by researchers have the same complaint. The National Science Foundation, the Lawrence Livermore National Laboratory, and many private companies are funding efforts to build speedier networks (transfer data at a hundred times the current speed of 10 million bits of data a second), which will change the nature of computing by making it possible for a scientist to receive live video or realistic color models on a workstation.[3]

Disadvantage

Inadequate standards like DXF, EDI, and IGES.

Comments

Though we have various data exchange file format standards, none of them are comprehensive enough to effectively exchange information among application programs used by PDIS. This has caused chaos in the industry when attempts are made to exchange information electronically. PDES and other standards committees are doing their best to address these issues. Now is the time for users of PDIS to come forward and work with potential vendors to define standardization and demand implementation of standards by the vendors.

After years of talking about Computer-Integrated-Manufacturing, the International Standards Organization (ISO) is considering a standard for shop-floor production and a CIM System Architecture Framework for Modeling that will set the stage for widespread vendor support. IBM plans to base the company's future CIM integration standard on the ISO model, known as CIM Open Systems Integration, or CIM-OSI. [2]

Disadvantage

Nonstandard and noncompatible mathematical representation of geometric entities.

Comments

B-Splines and other methods of managing geometric entities give rise to incompatible representation between the systems like CAD, CAE, and CAM. It is very important that all the systems used in PDIS are compatible. It will be good if all kinds of geometric representations are available in all the systems and are standardized. It may be a good idea for the contributor to present algorithms to standards so that everyone can offer all the variations of geometric entity representation to the standards body and eventually benefit users.

IMPLEMENTATION CASE HISTORY

Electro-Mechanical Product—Switch Gear

Electrical and Mechanical schematic information was directly transferred from the AutoCAD® drawings into the SYSTECH Product Definition System (SPDS). This effort saved a company 60% of effort compared to the manual methods. In simple terms, it took two weeks for the complete design, with an excellent quality, compared to the manual method which took six weeks on a previous similar design. Switch gear and panel drawings were made on AutoCAD; Terminal and Component tables were produced to check the accuracy of the drawings and design, and later Bill of Materials data were transferred to a centralized database where it was shared by other groups in need of information, such as estimating, marketing, and materials. Figure 5 shows the interface of SPDS to AutoCAD for switch gear product application. The part and drawing reference data were extracted from the CAD drawings and transferred to SPDS which maintains Bill of Materials and related data by the Interface program. Numerous reports for various departments were produced and some of the PDI was electronically transferred to other Manufacturing and Materials Systems.

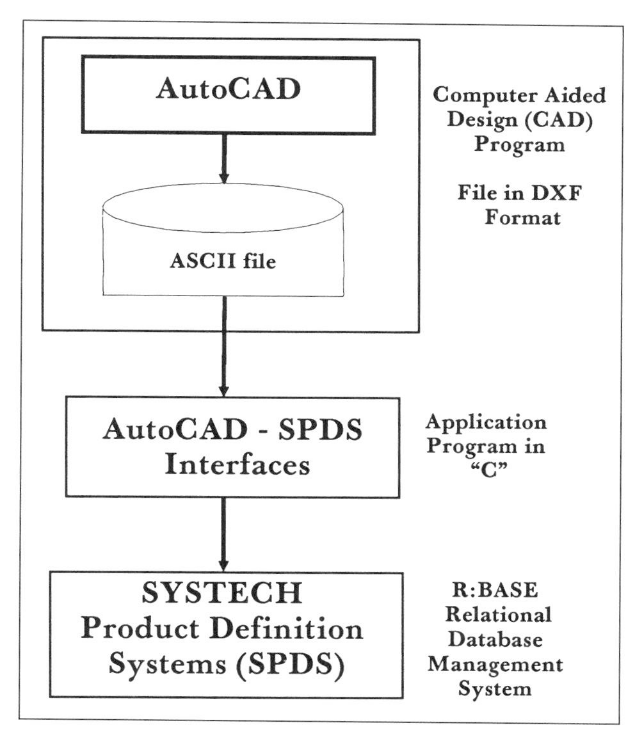

Figure 5. *AutoCAD-SPDS interfaces transfer product definition information on CAD to MRP and MRPII-type product, SPDS.*

Validation of Parts on CAD Drawings Against Approved Parts Database

All the releases of the drawings were maintained through the parts lists on the EIDS system developed on IMS/DB/DC. Any errors found while releasing the parts list were discovered and reported back to proper authorities and changes were initiated. The most common problem was the use of nonstandard or unauthorized parts found by checking against the approved parts databases maintained by the standards group. This was a major bottleneck and caused a great deal of delay. In need of a system to check for the accuracy of drawings and Parts lists, an interface (Figure 6) was developed to extract part numbers from the drawings on CAD on a computer at another site. Part numbers were sent to a computer connected by SNA network, where the authorized parts database was maintained in IMS, and fed back the results of the check to CAD stations where designers could correct errors. This validation check could be performed even while the designer was working on drawings prior to a formal release. On formal release, though not yet automated to feed into the EIDS database, a check of the parts was performed. This had tremendous savings in turnaround time to correct errors prior to release date and improve quality of drawings and parts lists.

Integrated Bill of Materials-Classic Example of Cooperation of Engineering and Manufacturing Departments

One logical Bill of Materials was designed for the whole company to manage all the PDI. This system had information transferred from CADAM drawings through Engineering Bill of Materials to Manufacturing Bill of Materials and subsystems. Both the as-designed and the as-planned product structures were maintained on the same database. Later, Manufacturing product structure data were transferred to Manufacturing database where other subsystems were linked to the Manufacturing Bill of Materials, such as Tool subsystem, Materials subsystem, and Scheduling subsystem. An attempt was made to link the Scheduling subsystem to the Engineering Bill of Materials. The system was logically designed but could not be exploited completely due to the software and hardware limitations of the installed system. This system served

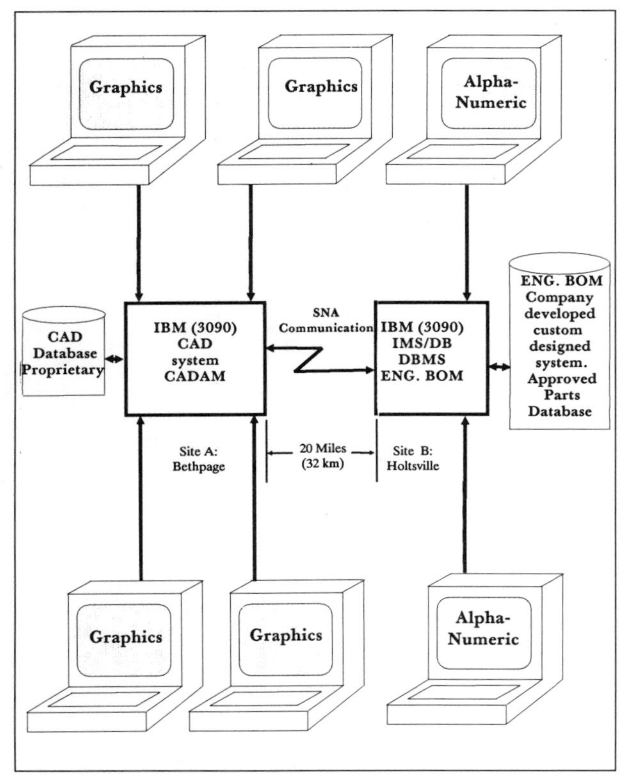

Figure 6. *Example of two different applications and their database management systems interfaced to exchange PDI.*

its purpose of providing all the necessary information for the engineering and manufacturing community.

Features, such as searching for the parts having a given thickness and listing its part and drawing numbers on which it exists to issue any change orders against the drawings, if the change in part thickness is required, and also providing logistics information when there is a stop order from the engineering on any part or assembly to track down all the affected parts of part or assembly being stopped, all helped to accomplish tasks which were impossible in the past. The manual system is inadequate to locate such information in a short time because all the drawings then have to be manually searched. The IBOM system was implemented on a network database system on IBM computers. Effective use of distributed databases and processors networked in a Client-Server Architecture system, like IBOM, can gain more mileage by providing information on-line at the appropriate workstations.

Release and Change Control of CAD Drawings

At one of the aerospace companies, software was written to manage Work-In-Process, Pre-Release and Release drawings, including change orders and revisions. Due to the file naming conventions, it was readily possible to use the file and application programs' drawing identification system, and use them to recognize the status of the drawings, including the date and time of release. Using this information, drawings were maintained on-line or archived based on the duration and need of the released information. The system was simple to design and implement, but the operation was still not streamlined and simple. Automatic transfer and availability of drawings on CAD to the user community and to the Engineering Product Structure databases was a plus and a time saver. With the least amount of personnel, it was possible to maintain the system. Manual data, though available and used, was being discouraged in lieu of making a good drawing and parts list on the Graphics system (CADAM).

CONCLUSION

With my experience in working with various systems to integrate PDI, the most challenging task is to overcome the resistance by supervisors and managers toward change and automation. They feel that they will lose control and will become dependent on computer systems, which they do not control and understand, in order to get results.

It is possible to network workstations and transfer information to where it is required from where it is produced or stored. So far, we have seen in most of the cases file transfer from one system to another, but now, selectively released product definition files and data can be transferred to the desired destination. Though most of the commercial software products' databases are proprietary, they still provide access to the product definition entities which include part numbers, 3D objects and Surfaces geometry, and other logical information like sizes and material properties. This access to proprietary databases is valuable in developing interfaces to extract desired PDI on a need basis.

This is the right time to mobilize and implement PDIS in the Engineering and Manufacturing companies. Technology will definitely drive automation, and redesign and rewriting of software is going to be normal day to day activity. In some instances, software is not yet available nor developed in-house. Updates and programming services must be a part of the ongoing system design. Either vendor, Systems Integrator, or in-house staff has to provide such services. There is tremendous opportunity to implement PDIS and reap the benefits for the company in optimizing availability of PDI by using computer systems. Computer Technology used carefully can be a great asset of any Engineering and Manufacturing company. Systems Integrators and operation engineers have very important roles to play in automating the factories to improve operation and cut down an order-to-deliver portion of the products' life cycles.

REFERENCES

1. G. Wiederhold, and R. Elmasri, "The structural model for database design." *Entity-relationship approach to systems analysis and design.* North Holland, 1980.

2. Paul E. Schindler, Jr., "Standards group considers architecture for CIM integration." CIM: Setting Standards, Systems, *Information Week*, July 16, 1990, p 46.

3. Zachary G. Pascal, "Making Networks as fast as machines they link. " *Technology*, Page B1, The Wall Street Journal, July 16, 1990.

BIBLIOGRAPHY

Carringer, R.A. and Birchfield, E.B., *Product Definition Data Interface: The link between islands of CIM automation.* (#MS85-122). SME Technical Paper, 1985.

Constantine, L.L. and Yourdon, E., *Structured Design.* Englewood Cliffs: Prentice Hall Inc., 1979.

Fulton, R.E., ed., *Managing Engineering Data: Emerging Issues.* New York: The American Society of Mechanical Engineers, 1988.

IGES, *Initial Graphics Exchange Specification* (IGES). Version 3.O. National Bureau of Standards, NBSIR 86-3359, 1986.

Integrated Information Support System (IISS), Information Modeling Manual, Extended IDEF 1X. Manhattan Beach: D. Appleton Company, Inc., 1985.

Leben, J. and Martin, J., *Data Communication Technology.* Englewood Cliffs: Prentice Hall, 1988.

PDES/STEP, *External Representation of Product Definition Data.* Working Draft of ISO TC184/SC4, 1988.

Rangan, R.M., Fulton, R.E., and Woolsey, T., EMTRIS: *Controlling Design and Manufacturing Information in a Discrete Part Manufacturing Environment.* Engineering Database Management: Leadership Key for the 90s, pp. 53-59, The ASME, 1989. (EMTRIS = Engineering and Manufacturing Technical Relational Information System.)

Rangan, R.M., Wheelan, P.T., and Fulton, R.E., "Managing Design/Manufacturing Information in a CAD/CAM Environment." *UPCAEDM '88*, Atlanta, pp. 228-235, 1988.

Rasdorf, W.J., and Fulton, R.E., eds., *Engineering Database Management: Leadership Key for the 90s.* New York: The American Society of Mechanical Engineers, 1989.

Sidle, T.W., "Weakness of Commercial Data Base Management Systems in Engineering Applications." *Proceedings of the 17th Automation Committee,* ACM, Minneapolis, 1980.

Teorey, T.J., Yang, D., and Fry, J.P., "A Logical Methodology for Relational Data Bases using the Extended Entity-Relationship Model." *Computing Surveys*, Vol. 18, No. 2, pp. 197-222, June 1986.

Yourdon, Edward, *Managing the Structured Techniques.* Englewood Cliffs: Prentice Hall Inc., 1979.

APPENDIX A

Abbreviations

COMMUNICATION

EISA	Extended Industry Standard Architect
Ethernet	LAN communication protocol
FDDI	Fiber Distributed Data Interface
ISDN	Integrated Services Digital Network
LAN	Local Area Network
MAN	Metropolitan Area Network
MAP	Manufacturing Application Protocol
MODEM	Modulation and Demodulation—for communication
OSI	Open Systems Interconnection
RF	Radio Frequency
RS232	Serial port communication standard
SCSI	Small Computer System Interface
SNA	Systems Network Architecture
TCP/IP	Transmission Control Program/Internet Program
TOP	Technical Office Protocol
WAN	Wide Area Network

COMPUTER INTEGRATED MANUFACTURING

2-D	Two Dimensional Drafting
2-1/2-D	Two and One-half Axes Drafting

3D	Three Dimensional Design
CAD	Computer Aided Drafting
	Computer Aided Design
CAD/CAM	Computer Aided Design/Computer Aided Manufacturing
CADD	Computer Aided Design and Drafting
CAE	Computer Aided Engineering
CAM	Computer Aided Manufacturing
CAPP	Computer Aided Process Planning
CASE	Computer Aided Software Engineering
CAT	Computer Aided Testing
CIM	Computer Integrated Manufacturing
CNC	Computer Controlled Numerical Control
DNC	Direct Numerical Control
EIS	Enterprise Information System
FEA	Finite Element Analysis
FEM	Finite Element Modelling
	Finite Element Method
JIT	Just In Time
MCAE	Mechanical Computer Aided Engineering
MRP	Materials Requirements Planning
RPII	Manufacturing Resource Planning
NC	Numerical Control

DATABASE MANAGEMENT SYSTEMS

DBMS	Data Base Management System
IMS/DB/DC	Information Management System/Data Base/Data Communication
OODBMS	Object Oriented Data Base Management System
RDBMS	Relational Data Base Management System
SQL	Structured Query Language

MANUFACTURING APPLICATION SYSTEMS

COPICS	Communication Oriented Production Information and Control System
GDT	Geometric Dimensioning and Tolerancing
SCADA	Supervisory Control And Data Acquisition
SFC	Shop Floor Control
SPC	Statistical Process Control

MISCELLANEOUS

CADAM	Computer Aided Design And Manufacturing
CD-ROM	Compact Disk-Read Only Module
CIM-OSI	Computer Integrated Manufacturing-Open Systems Integration
COTS	Commercial Off The Shelf products
DEC	Digital Equipment Corporation
DOD	Department Of Defense
EIDS	End Item Definition System
E-MAIL	Electronic Mail
Esprit	A Consortium of 21 European Computer Vendors (including IBM)
GUI	Graphics User Interface
HIPO	Hierarchical Input Processing Output
IBM	International Business Machines
ICAM	Integrated Computer Aided Manufacturing
IDEF0	Modeling Methodology developed by ICAM Project of Air Force Manufacturing Technology
ISO	International Standards Organization
NURBS	Non-Uniform Rational B-Spline
PC	Personal Computer running under MS-DOS
PDIS	Product Definition Information System (a kind of generic system name which manages Product Definition Information, like CAD, CAM)
SPDS	SYSTECH Product Definition System
WORM	Write Once Read Many (Compact Disk)

OPERATING SYSTEMS

DOS	Disk Operating System
MVS/XA	Multiple Virtual System/Xtended Architecture—an operating system for IBM mainframe computers
UNIX	AT&T's Multi User Operating System
VMS	Virtual Memory Systems—an operating system for VAX computers

PRODUCT DATA INTERFACES

DXF	Data Exchange Format
EDI	Electronic Design Interchange
EDIF	Electronic Design Interchange Format
IGES	Initial Graphics Exchange Specification
PDES	Product Data Exchange Specification

PROJECT MANAGEMENT SYSTEMS

CPM	Critical Path Method
GANTT	Bar Chart, primarily to control the time element
PERT	Program Evaluation and Review Technique

Presented at the CASA/SME AUTOFACT '89 Conference, October 30-November 7, 1989

Distributed Databases in a Heterogeneous Computing Environment

LAWRENCE A. ROWE
University of California at Berkeley

This paper describes alternative software architectures for database applications. The ANSI SQL standard data model and the database vendors commitment to tools and systems that support portable applications will allow organizations to change hardware platforms and reconfigure their computing environment more easily. The features of a full-function distributed database system and how they can be used to solve some common problems in a large manufacturing organization are also described. Lastly, database gateways are described that interface existing databases stored in older database systems (e.g., hierarchical and network) to applications that use the relational model.

INTRODUCTION

Engineering design and manufacturing environments are heterogeneous computing environments. A typical environment will have many different types of computers (e.g., embedded computers, personal computers, workstation computers, minicomputers, and mainframe computers) that are interconnected by a network. Many different applications including engineering design, real-time factory control, and corporate business applications are run on this collection of computers.

These applications manage databases that contain relevant data. For example, engineering design applications maintain documents that specify a product's requirements and design, geometric models that describe a product, and component lists that specify the parts that make up a product. A factory control system keeps track of the work-in-progress (WIP), status of equipment in the factory, and test data collected during the manufacturing process. Finally, corporate business applications manage data about orders, personnel, and inventory. These information systems (i.e., the application programs, the databases, and the operational procedures) are an important corporate asset.

Computing environments of the past were dominated by a few large mainframes. The proliferation of low cost microprocessors has led to the introduction of hundreds, and in some cases thousands, of different types of computers into the engineering design and manufacturing environment. Application programs must run on these different computers. At the same time, they must access a shared, integrated database created by merging the autonomous application-specific databases. A shared, integrated database is required for two reasons. First, applications will have to access data in more than one database to meet future business requirements. And second, existing applications that currently run on a single large computer will be distributed to different computers in the environment.

An important trend in the development of application software is to build portable, reconfigurable applications. A portable application can be run on different hardware platforms and different software systems (e.g., operating systems and database systems). A reconfigurable application can be distributed across different computers in different ways. Vendor independence is important because it allows an organization to change computers without having to rewrite the application software. Consequently, the organization can migrate to more cost effective computer systems with less difficulty. Application reconfiguration is important because the distribution of the software across different computers may change due to changes in the price performance of hardware.

This paper surveys the current state of database systems and database application architectures. Alternative software architectures and database system support for those architectures will be presented. We will attempt to show that a standard data model and a vendor commitment to portable applications allows an organization to build portable, reconfigurable applications.

The software environment we envision is applications written in a 4th Generation Language (4GL) or a 3rd Generation Language (3GL) that access a relational database using the SQL query language.[1] These applications may interact with the user through an alphanumeric or graphic terminal and they may run on different operating systems.[2]

The remainder of the paper is organized as follows. Section 2 discusses alternative software architectures for applications and the DBMS. Section 3 presents the features that a full-function distributed database system should provide. And, Section 4 describes database system functionality that is required by engineering design and manufacturing applications.

SOFTWARE ARCHITECTURES

This section describes software architectures for applications and database management systems (DBMS). These architectures can be used in different ways in an engineering design and manufacturing environment. Some examples are discussed at the end of this section.

The application program must run in a separate process from the DBMS as shown in Figure 1. The application process contains codes specific to the application and to interact with the user through a terminal or workstation interface. The DBMS process contains the database system code. The DBMS is run in a separate process so that it can run in a different protection domain. That way, the application program can access only the data that it is supposed to access.

Another reason the DBMS is run in a separate process is so that it can be run on a different computer. This configuration is sometimes called remote database access. The advantage of this configuration is that the workload can be spread across two computers

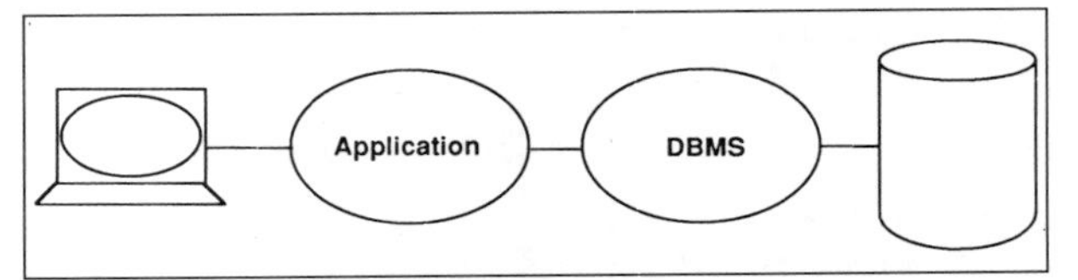

Figure 1. Separate application and DBMS architecture.

rather than just one computer. If the computer that runs the DBMS process is dedicated to that function, it is called a database server. This configuration is a natural one when users are running applications on a workstation but still need to share a common database. The applications access the same database managed by the DBMS process on the database server. This configuration has the added benefit of putting the DBMS code close to the disk that contains the data and the interface code close to the user. Because the application issues high-level relational queries to the database server, only the data required by the application must be sent to the workstation. In contrast, if the DBMS process ran on the workstation and accessed the data through a network file system, all disk pages that the DBMS read would have to be sent to the workstation. A similar argument can be made that the application process should be located close to the user to minimize the messages that must be sent between the interface devices and application program.

In a multiuser environment, each application might have a separate DBMS process as shown in Figure 2. The DBMS processes run in "shared instruction" mode so that only one copy of the DBMS code must be kept in main memory. However, this architecture will be less efficient than the server architecture shown in Figure 3. The server DBMS process handles requests from more than one application.

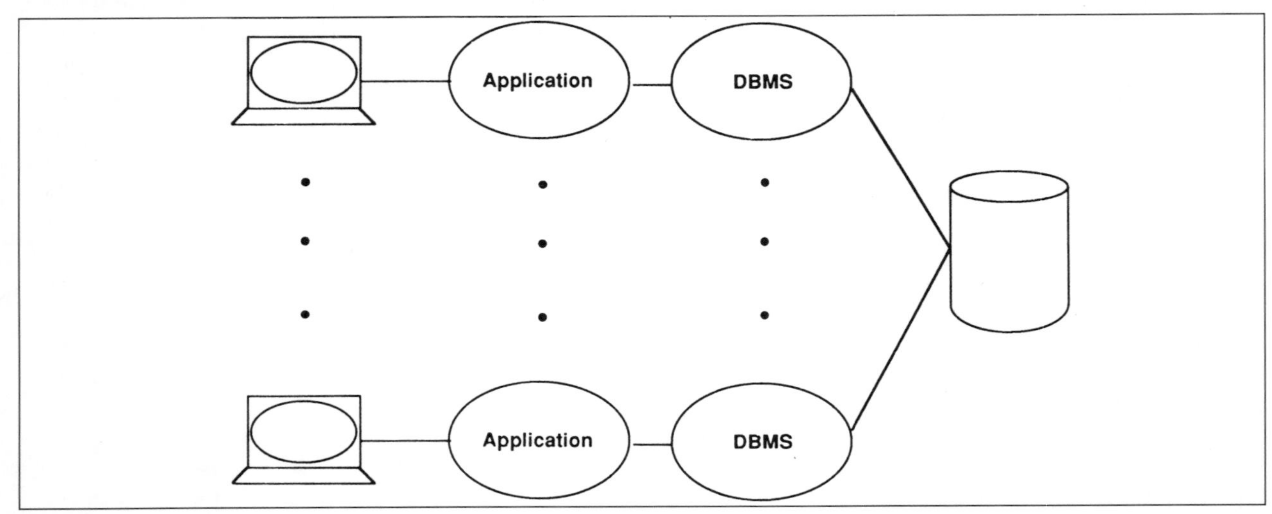

Figure 2. Process-per-user configuration.

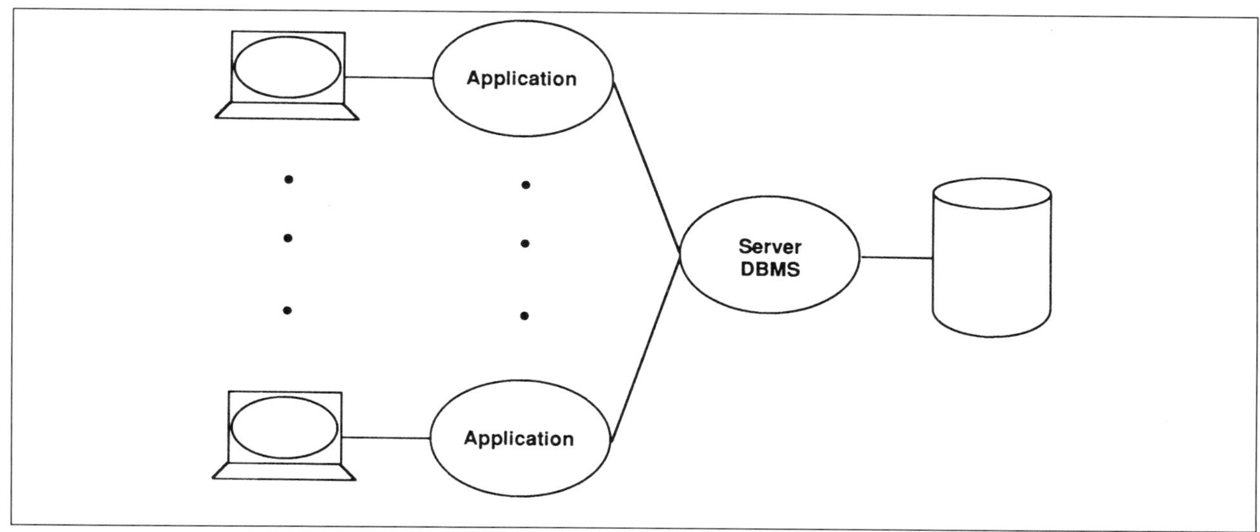

Figure 3. *Server DBMS architectures.*

The server DBMS is a mini operating system that manages and schedules "tasks" (i.e., the application queries), issues I/O requests to read and write disk pages, and manages a buffer pool of recently accessed pages.

The primary advantage of a server DBMS over a process-per-user DBMS is that the server can be customized to the database tasks that it is executing. A customized operating system is more efficient than a general-purpose operating system because it knows more about the tasks that it is running.[3] For example, the server has a good idea of how much CPU time and how many I/O's each task will perform because the query optimizer estimates these numbers when it selects a query execution plan for the task. Most on-line transaction processing systems use a server architecture because it is the most cost-effective architecture.

A server architecture is a good architecture, but it does not take advantage of the tightly-coupled parallel processors that are currently popular. The appropriate configuration in that environment is a multiserver architecture. Figure 4 shows a multiserver architecture. Each server runs in parallel on a different processor. The advantage of this architecture is that both response time and output can be improved because the system can execute many tasks in parallel. More work will be completed by the multiserver architecture that exploits parallelism than by a server architecture that uses interleaved execution. Another advantage of the multiserver architecture is that DBMS cycles can be increased by adding another processor if the application load grows to where it cannot be handled by an existing computer system.

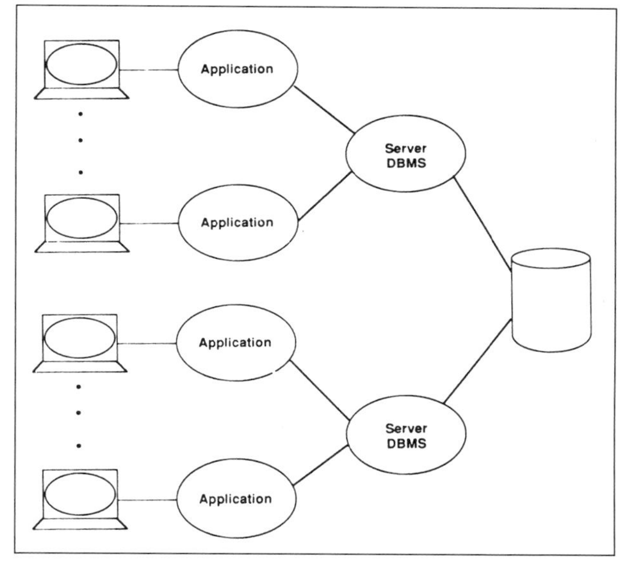

Figure 4. *Multiserver architecture.*

Parallel processors are most frequently used to support a larger application workload. However, some vendors are developing query executors that will use several processors in parallel to speed up query execution. Parallel execution is also possible in a distributed database system as discussed below.

Thus far, we have focused on applying the idea of servers to the DBMS program. The same idea can be applied to the application program. Figure 5 shows an application server process that is connected to several database servers. The application server can send a request to the server process that is least busy in order to balance the load across the different processors. A disadvantage of this architecture is that the application server can become a bottleneck which will limit the system performance. Consequently, it is necessary to support multiple application servers. Although some systems use this architecture, it conflicts with the trend of moving applications to a user's workstation.

Some organizations cannot justify the cost to put a high-priced graphic interface workstation on every person's work space. Some window systems, notably the X Window System[4] and Sun's NeWS System,[5] actually run a server process that performs screen output and handles keyboard and mouse input.[6] The application process architecture in the workstation is shown in Figure 6. Each application interacts with multiple windows that are managed by the window system server. A new generation of intelligent terminals that execute just the X Window System server are being developed that will cost roughly $1,000 and run exactly the same applications with the same interface "look and feel" as a more expensive workstation. The advantage these devices offer is that a low cost-per-user graphic terminal can be used that will run exactly the same applications as a high cost-per-user workstation. The X terminal solution may run slower than the workstation for some applications because the application runs in a shared compute server rather than in the local workstation.

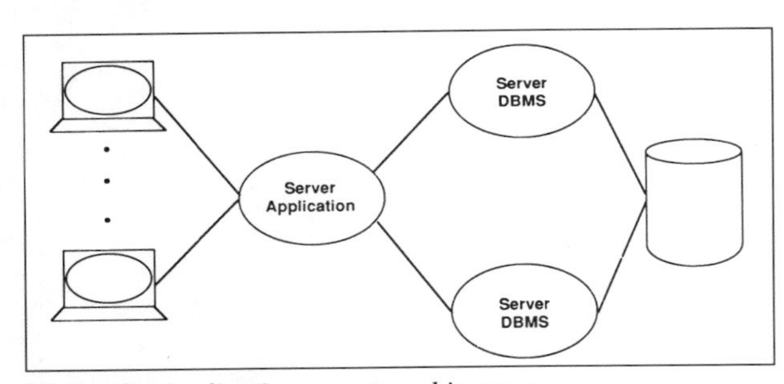

Figure 5. Application server architecture.

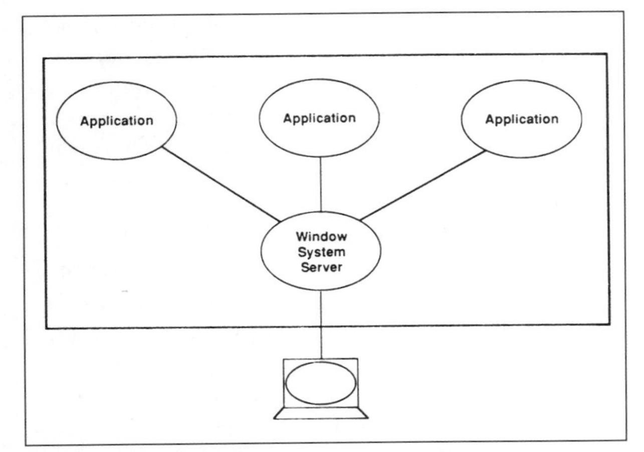

Figure 6. Workstation window server architecture.

The last database system architecture that will be discussed evolved from the desire to access data stored on physically separate computers. This architecture, called a distributed database system, is shown in Figure 7.[7] The FE application sends exactly the same commands to the distributed DBMS that it sends to a single-site DBMS. The distributed DBMS sends commands to the local DBMS's to implement the application query. Typically, the application process and the distributed DBMS process run on the same computer and the local DBMS's run on the computers that contain the data. This architecture has three advantages. First, data can be accessed transparently. That is, an application program can access data stored on different computers without having to know where it is stored. Second, queries can be executed in parallel to improve performance. And third, computers can be incrementally added to or removed from the system without requiring changes in the application programs. The next section describes in more detail the features of a distributed database system.

Figure 8 shows a typical hardware configuration in a factory. A local area network connects together a factory computer, a database server, and a collection of workstation and cell computers. In addition, terminals are connected to the factory computer and to

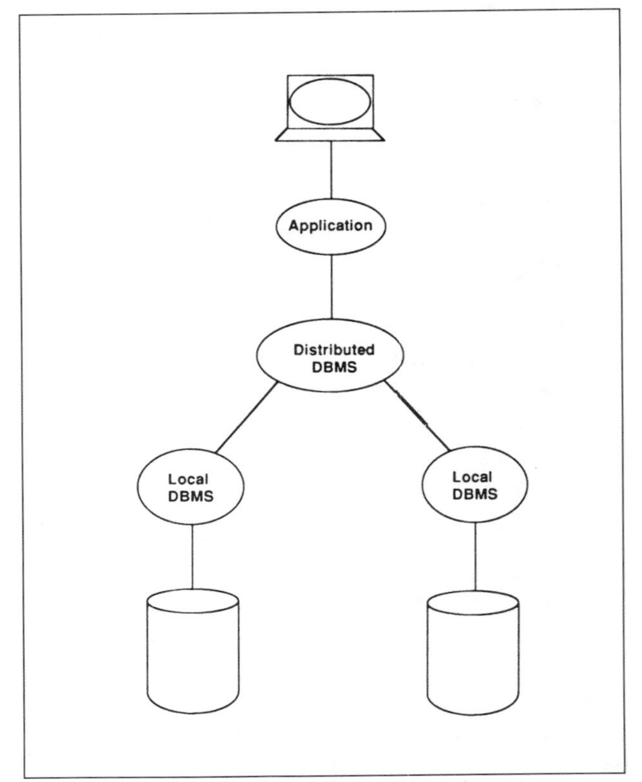

Figure 7. Distributed database system architecture.

some cell computers. Notice that the network has gateways to the corporate and engineering design computers too. The data for the WIP system that controls the manufacturing process will be distributed to the database server and the cell computers. Most of the time the applications that run locally on the cell computers will access data on the local database and occasionally they will access data on the database server. A server DBMS should be run on the cell computer that can handle the local applications and the commands sent to it by a distributed database process that runs on the factory computer or one of the workstations. An engineer troubleshooting a problem in the factory might query data in the database server or a local database on a cell computer. However, all these applications should be written in an SQL compatible language so that they can be easily ported to run on the different databases.

This section described alternative architectures for database and application systems. A typical factory computing environment was described and ways to use the alternative database architectures in that environment presented. Most factories have heterogeneous hardware so it is crucial that the application software run on this different hardware without changes.

DISTRIBUTED DATABASE SYSTEMS

Distributed DBMS's have only recently been introduced to the commercial marketplace. Consequently, most products are primitive. This section describes the features that a full-function distributed DBMS should eventually support. Current products will be enhanced to provide most, if not all, of these functions over the next five years.

First and foremost, a distributed DBMS must be functionally equivalent to a single-site DBMS. It must support *ad hoc* and program query access to the database, transactions (i.e., multiple user access and crash recovery), integrity and protection, and other single-site DBMS services. These functions must transparently operate on data located at different sites.

A distributed DBMS can provide more function than a single-site DBMS because it is distributed. The natural partitioning of a database is to place different tables at different sites. Queries that involve one table can be executed at one site. Queries that involve tables at more than one site can either move all the data to one site and execute the query there, or move subsets of the data to several sites and execute the query at those sites. The distributed query optimizer is responsible for picking an efficient query execution plan given a query, a particular data distribution, and information about the cost to move data and execute local queries.

A table can be distributed to more than one site by either vertical or horizontal partitioning. A vertical partition stores different columns of the table at different sites. For example, given the employee table

EMP (Name, Address, Dept, Picture)

The Name, Address, and Dept columns could be stored at one site and the Picture column, presumably a large pixmap, could be stored at a different site. A horizontal partition stores rows of the table at different sites. Each partition is called a *fragment*. The EMP table could be partitioned based on the employee's department. For example, administrative employees could be stored at headquarters on the corporate computer and manufacturing employees could be stored in the factory computer in the manufacturing plant. Regardless of how the data is partitioned, the following query should return the same results:

```
Select   *
from     EMP
where    Name = 'John Smith'
```

The data partition is chosen to optimize a particular set of queries. For example, if employee pictures are infrequently accessed they can be stored on a slow optical disk in the corporate data center. This example uses vertical partitioning to store infrequently used data on slower devices. On the other hand, suppose there is a very large table, say 10 gigabytes, of historical sales data that a marketing person is analyzing for trends. This table can be horizontally partitioned and distributed to different computers so that the *ad hoc* queries run by the marketing person can be executed in parallel to improve response time.[8]

A distributed database can be accessed from geographically dispersed places. For example, corporate headquarters might be in Detroit but a manufacturing plant might be in Mexico. Troubleshooters in the manufacturing plant might need to access a part database stored at headquarters. If many queries against the part database are being run, the data might be moved many times to Mexico. It may be impractical to move the part database to Mexico permanently because it is also being accessed by other plants in the US. The solution might be to keep a copy of the data in Detroit and Mexico. The distributed DBMS will manage the copy. In other words, the distributed DBMS will choose the least expensive copy to access and it will

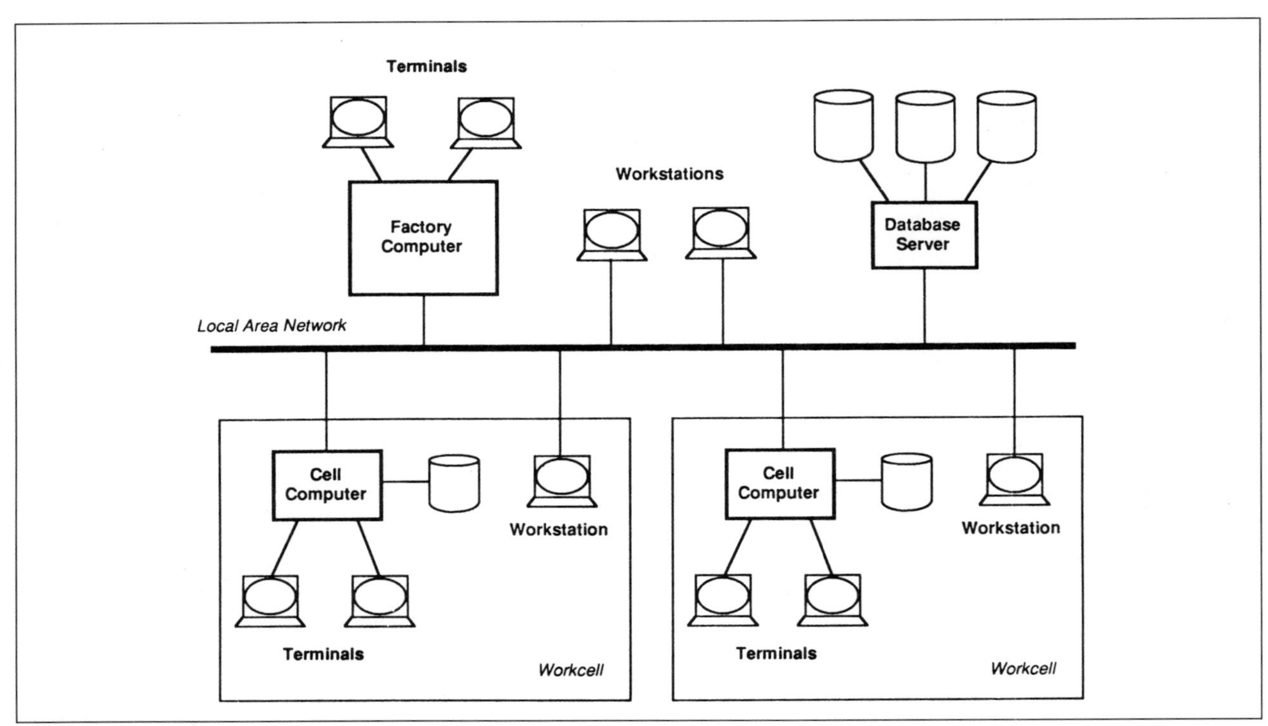

Figure 8. *Typical computing environment at a factory.*

propagate updates to all copies to maintain database consistency.

Copies are a good idea if you have data that does not change frequently. Otherwise, the cost of maintaining the copies may outweigh the advantage of keeping it. Another possibility is to maintain a snapshot of the part table in Mexico. A snapshot is just a copy of the table at a particular time. Then, the snapshot can be updated with changes at regular intervals (e.g., once a day, week, or month). A snapshot improves the execution of queries in Mexico, but the answers may be out of date. This solution might be perfectly acceptable if the parts table does not change very often. The distributed DBMS should manage the periodic updating of the snapshots.

Another feature of a full-function distributed DBMS is replicated catalogs. A catalog is a table maintained by the DBMS to keep track of the database itself. For example, a catalog exists that describes each table or fragment in the database and the site at which it is stored. A single catalog leaves the distributed DBMS vulnerable to a single failure. The entire database will be unavailable if the site that holds the catalog is down. The solution is to replicate the catalogs at several sites. That way, if one site goes down, the distributed DBMS can access a copy of the catalog at a different site. Replicated catalogs are not free. Schema changes (e.g., adding or removing indices, adding columns to a table, or adding or removing tables) will require that all copies of the catalogs be updated. Consequently, replicated catalogs will be useful for stable production environments that need high availability.

The last feature of a distributed DBMS is a distributed transaction log. Recall that log records have to be written before a transaction can commit so that the database can be restored to a consistent state should the system crash before the updated data pages are written to disk. A limited function distributed DBMS will have a log at one site. This solution is acceptable if the overhead to write the log is small as will likely be the case if the log is at the same geographical site. However, if the log is at a remote site, it might take a long time to write the log. The solution to this problem is to have many logs distributed at different sites. Distributed logs do not cause a problem for the applications that run on a distributed DBMS, but they do introduce considerable code complexity into the distributed DBMS itself.

HETEROGENEOUS DISTRIBUTED DATABASE SYSTEMS

Distributed database systems solve the problem of building integrated information systems that use geographically dispersed databases and parallel processing. They do not, however, solve the problem of managing the transition from existing applications and databases to these new systems nor the problem of integrating disparate data that is not stored in a database (e.g., geometric models stored in a file). A distributed database system can be used to solve these problems if we build gateways from the distributed system to the older systems.

Figure 9 shows a heterogeneous distributed database system that interfaces to an IBM IMS database and a DEC RMS file. The IMS and RMS gateways translate the commands sent by the distributed DBMS process to a local database system into commands on the foreign database or file system. The results of executing these commands are translated into responses to the DBMS process which passes them back to the application program.

The majority of relational queries can be translated to a foreign database. However, some queries, usually update commands, cannot be translated. For example, a relational database with EMP and DEPT tables that represent employees and departments can be updated so that employees exist who are not in any department.[9] This update cannot be mapped to a CODASYL system that represents the departments by DEPT records and places the employees in owner-coupled sets.

Other problems must be solved to make these systems practical. For example, a common dialect of SQL must be adopted so that gateways between different relational databases can be implemented. Common data types, catalogs, and error messages must also be specified so the systems can work together smoothly. Nevertheless, heterogeneous distributed databases can be built and they will solve an important problem.

SUMMARY

Single site relational database systems are approaching maturity. Functions can still be added to the leading products, but most products already have a rich set of functions. Distributed database systems are an important new technology that will allow data stored at different sites to be accessed and used as though it were at a single site. Distributed DBMS's are immature products. However, they offer real promise for several problems. First, they will allow applications written against a relational database to access data stored in different physical databases. The data can be stored at different sites and in different DBMS's (e.g., hierarchical, network, or relational).

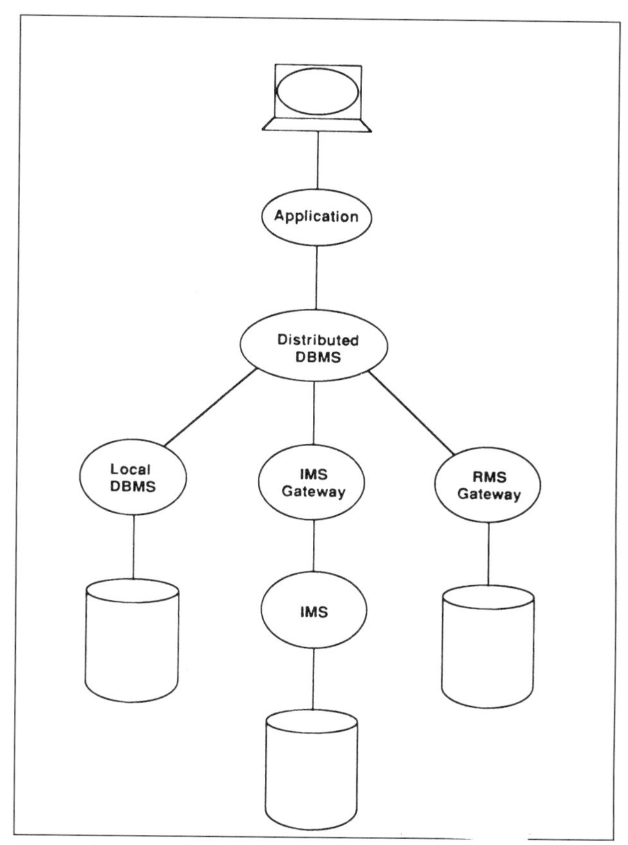

Figure 9. Heterogeneous distributed database system.

Second, they will take advantage of the proliferation of low cost computers. Parallel processing can be exploited to handle changing workloads and to improve response time for selected queries.

These capabilities will be available without changes to application programs because a standard program interface to a database is used and because database vendors recognize that customers want portable applications that can access heterogeneous databases.

It may take five years for the vendors to complete the network protocols, database gateways, and distributed database systems. Nevertheless, the direction for the future is portable, reconfigurable applications that run on heterogeneous computing environments.

REFERENCES

1. C. J. Date, *An Introduction to Database Systems,* Addison-Wesley, Reading, MA, 1981.

2. The goal of portable applications will be easier to achieve if a UNIX operating system is used. However, many 4GLs and 3GLs are designed to run on different operating systems.

3. M. Stonebraker, "Operating System Support for Database Management," *Comm. of the ACM*, July 1981.

4. R. W. Scheifler and J. Gettys, "The X Window System," *ACM Trans. on Graphics* 5, 2 (Apr. 1986).

5. *NeWS Programmer Guide*, Sun Microsystems, Inc., Mar. 1989.

6. A protocol could be defined to run the other major window systems (Microsoft's Presentation Manager and the Macintosh Toolbox) as servers, but they do not inherently support a window server.

7. S. Ceri and G. Pelagatti, *Distributed Databases: Principles and Systems*, McGraw-Hill, New York, NY, 1984.

8. A simple query to compute a mean that takes 48 hours to run on a single processor can be run in 60 minutes on a parallel processor.

9. This example assumes that the database does not have a referential integrity constraint that disallows this situation.

This rcsearch was supported by National Science Foundation Grant MIP-8715557.

Reprinted from Manufacturing Engineering, May 1988

Which Network is The Right One?

JAMES G. AMES
Arizona State University

INTRODUCTION

Information is widely recognized as a vital resource. There seems to be a well traveled path between awareness and implementation, however. Companies generally automate their financial information first, followed by their administrative and engineering information. Eventually, the value of accurate, coordinated, and timely manufacturing information is appreciated, if not yet attained. Usually, automated information systems are implemented along departmental lines. Intersystem communication, when it happens, generally takes place via human intervention either by paper or by direct contact. The nirvana of information resource management is an information utility that can be accessed by users or providers of information regardless of location or function, much like an electric utility (Figure 1).

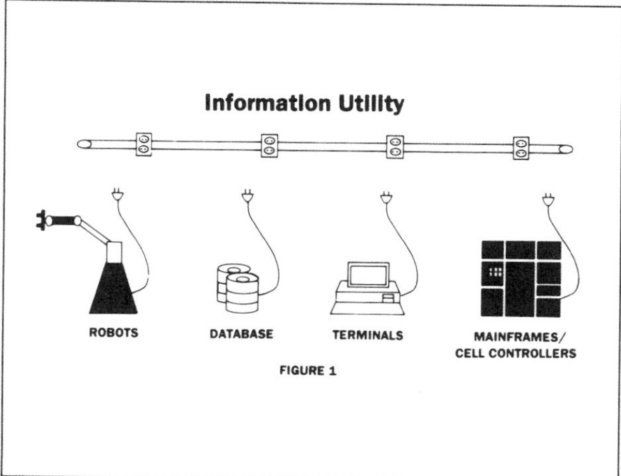

Figure 1.

Four technological, and one organizational, requirements are central to the realization of automated information management. The technologies are the following:

- Computers are at the heart of any automated system. While the concepts are well understood, rapid advancement of computer technology, coupled with a much wider spectrum of performance capabilities, makes system integration issues complex.

- Database design and implementation issues must be addressed if the Babel complex of private data reserves is to be breached. Considerable progress has been achieved recently in research and implementation of integrated distributed databases.

- Network technologies, which allow the seamless integration of data and processes residing on a number of computers, are essential. To be truly effective, this integration should include computers of various vendors and capabilities.

- People with the training and motivation to excel in an automated environment are vital, and—because humans cannot be redesigned—it is important that equipment be easy to use.

The other requirement for any successful automated information system is that it be consistent with the organization's goals, strategic plans, and structure. Like any other tool, an information system should facilitate the efficient production of a company's products.

Too often, management looks at an information system as a way to force changes in organizational structure. For example, corporate finance could gain control of decentralized production units by requiring that all billing be done via a finance controlled computer with remote data entry. This may or may not be advantageous to the company's efficiency and bottom line. If changes are warranted, they should be reviewed and reflected in the overall strategy and structure before an information system is implemented or changed.

Potentially more harmful, however, is the fact that the effects of implementing an information system may be neither planned nor understood. Many managers seem to believe that installing state-of-the-art equipment will, in and of itself, lead to a factory that is competitive in today's world marketplace. In some organizations, new technology can be detrimental. Employees accustomed to coordinating their actions with a one-sentence verbal request may not take kindly to paging through a three-screen menu on a computer terminal.

CIM STRATEGY

The introduction of computer-integrated manufacturing (CIM) in the context of enterprise-wide information management should be accompanied by a complete reevaluation of the company's strategic plans and operational parameters. For a surprising number of firms, these vital guidelines may not currently exist except as the traditional "That's the way it's always been done" corporate culture.

As understanding of automated information system technology grows, "new" organizational strategies seem to be emerging based on the premise that integrated information management facilitates decentralized decision control. Competitive manufacturing requires rapid response and maximum utilization of human and other resources while maintaining corporate control of strategic organizational goals and objectives. Research and pilot projects are exploring the owner/operator or intelligent foreman's workstation (Figure 2) in automated production units. In a way, CIM based on integrated information systems makes possible organizational structures not seen since the pre-Frederick W. Taylor days of skilled craftsmen.

A coordinated CIM program requires an effective communication selected on the following criteria:

Business Strategy

Networks utilized in a company's facilities must be consistent with organizational strategy. For example, if quality is deemed to be of paramount importance, the quality assurance department may be given access to all production and scrap information and the ability to immediately forward high-priority status alerts to plant management or shut down the assembly process.

Flexibility

The maxim that the only constant is change is especially true in modern manufacturing. General Motors has reported as part of its justification of MAP (Manufacturing Automation Protocol) networks that

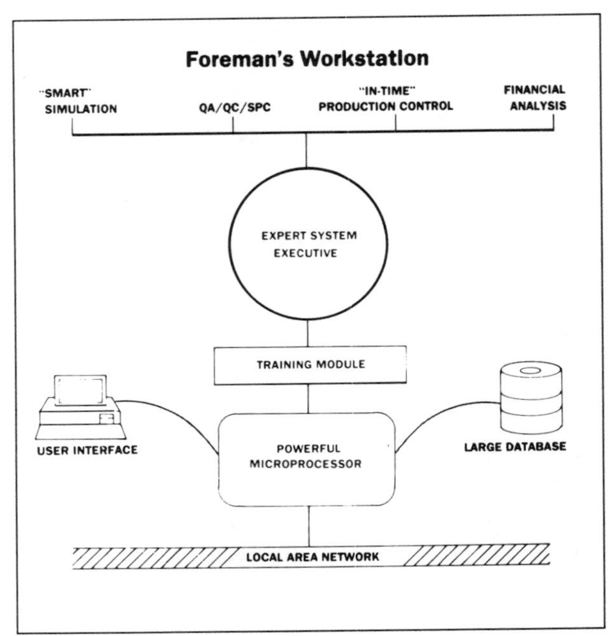

Figure 2.

more than 50% of new systems costs were directly attributable to communications. (1) The best strategic planning can generally only predict trends and rates of change. It is seldom possible to anticipate very far in advance which specific tools will be needed or available.

Capacity Required

Control and data flow requirements can be modeled utilizing tools such as IDEF0 and IDEF1 from the U.S. Air Force ICAM project, and basic network performance models can be constructed with tools such as CACI Product Co.'s (La Jolla, CA) NETWORK II.5. In the final analysis, however, acceptable network output and response times are dependent on a high quality network design team with an extremely good understanding of the organization's current characteristics and goals. Of course, any capacity plan will include data-flow-per-unit time and response time requirements for each node on the system. Capacity planning, however, should also include anticipated growth strategies consistent with the strategic plan.

Environmental Conditions

Every aspect of an information environment must be considered. For example, considerable electromagnetic interference (EMI) might dictate specially shielded coaxial transmission cables. The presence of a mouse population fond of eating cable insulation may indicate the advisability of installing steel conduit. Workers with heavy gloves and dark goggles may need one type of terminal, while those in an air-conditioned office might require a different configuration.

Economics

Life-cycle costs should be understood and accurately predicted for any information system before it is constructed. Return on investment, however, is not easy to predict with normal accounting methods. The value of an integrated information system must be in the context of all the corporate functions. Some say the lowest-cost network is one that performs all required functions, transparently uses existing user, application software, never fails, is easily expanded to provide for the seamless integration of new functions, manages itself for optimum performance, requires no training to operate, and costs $1. Wish lists like this are useful for determining which features are absolutely essential and which can be relaxed, and by how much. The result should be acceptable performance at the lowest cost, tailored to the company's needs.

Common Sense

Communication in a factory environment can be accomplished in a variety of ways, occasionally by two or more methods at the same time. For instance, communication can take place via digital electronic networks, hand-carried written messages, verbal conversations, or even visual cues. An effective network plan will evaluate all possibilities and select the most appropriate. Product marketing may be interested in monthly sales, but real-time production scheduling data would be inappropriate and costly. How well a local area network (LAN) works in a business environment is a direct function of its designers' understanding of communication requirements and their ability to develop solutions that accommodate existing operations while achieving planned improvements. Research work now under way may lead to quantitative/qualitative modeling of existing and proposed communications (Figure 3), but, at this time, a great deal depends on the intuition of the network designers.

INTEGRATING INFORMATION SYSTEMS

An integrated information system plan can include open interconnect strategies, closed or vendor-specific implementations, or a mixture of the two. An open system is one in which the operating parameters of hardware and software used to exchange information are defined by recognized standards that are complete and testable. Open systems must be interoperable in that any device conforming to defined standards can communicate with any device from any vendor that uses the same standards.

The International Standards Organization (ISO) Open Systems Interconnect (OSI) seven-layer model provides a well-recognized architecture for an open communication system. When specific standards are written and accepted for each layer of the communication protocol, hardware can be constructed and applications software written that interoperate regardless of vendor. MAP is a good example of the OSI model. It is possible, however, to have an open protocol such as TCP/IP (Transport Control Protocol/ Internet Protocol), which has been mandated by the Department of Defense but does not conform directly to the OSI model. With a little forcing, however, TCP/ IP can be viewed as a subset of the OSI model.

For all practical purposes, efforts on "open" LANs are concentrated in three closely related areas: MAP, TOP (Technical and Office Protocol), and GOSIP (Government Open Systems Interconnect Profile). At the present time, GOSIP-compliant networks may utilize the protocol stacks of either MAP, TOP, or TCP/IP.

Closed communications systems are based on proprietary vendor solutions and, in general, will not operate with any other type of equipment. Between black and white, however, there are many shades of gray. Vendors can claim openness for their protocols by publishing the procedures and not suing competitors who use them. The outstanding example of a standard developed by imitation is the IBM PC/XT/AT. This is not likely to happen often, however, and in complex areas like networking may not provide an acceptable universal solution in any event.

The user can combine system types via gateways (devices designed to translate messages from one format to another). For example, a company with IBM business computers and Digital Equipment factory computers may choose SNA-DECNET Gateways as the integration tool. The variety of combinations is seemingly endless. In general, however, it is very difficult to change multivendor systems with heterogeneous protocols without affecting major operational features.

As is apparent, open systems network architectures make good sense from a CIM planning point of view. In the long term, the need for information system flexibility and interoperability cannot be overempha-

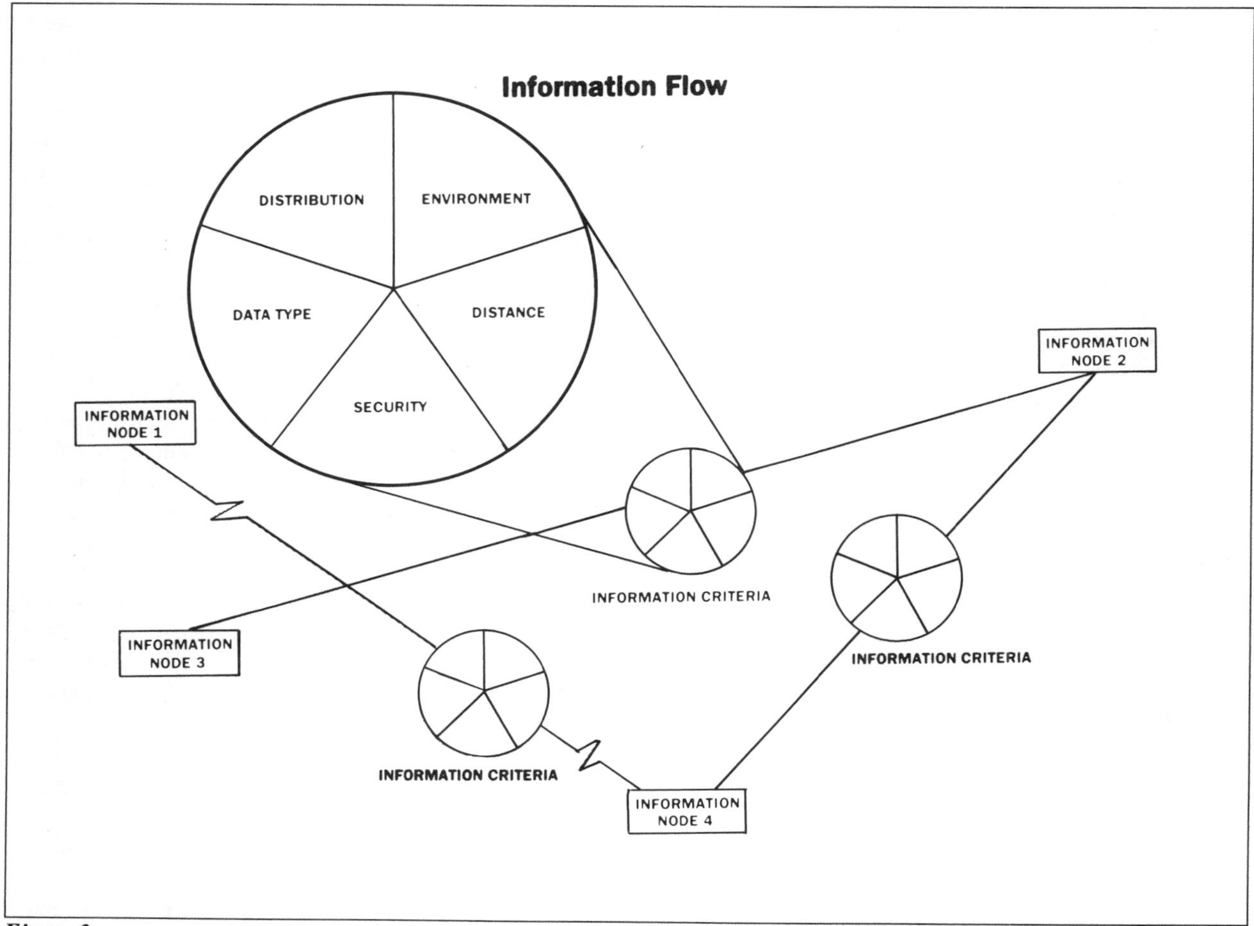

Figure 3.

sized. There are still very good reasons, however, for using proprietary networks.

An open system is inherently more flexible in some respects. It is relatively simple to connect MAP hardware from several vendors on the same network and get them to exchange data reliably. But problems develop in network management. No easy-to-use, effective management capability operates with all vendor equipment. Applications interfaces to user software are not well defined, and database access is not standardized. The good news is that once a company has developed applications software, it can be reused many times in spite of numerous network changes. The bad news is that small companies will have difficulty justifying large initial development costs. Costs for user-developed applications software have been estimated at 60% of network expenditures. These problems will diminish or disappear as the use of MAP/TOP spreads.

Growing acceptance of MAP/TOP is expected in the next few years. This will occur for several reasons:

- The specifications are maturing, and no major revisions will be made to MAP/TOP 3.0 for six years.
- Products based on MAP/TOP 3.0 will be more available and useful as vendors respond to major customers.
- International interest and acceptance of standards will make open systems attractive to multinational companies.
- The influence of the U.S. government requiring GOSIP-compliant networks for defense contractors will be an important impetus to open systems products.

Even with increasing availability of OSI networks, vendor proprietary systems will be the best fit for some areas for a long time. There are several reasons why a company may want to use a closed system:

- A system vendor may be able to provide a total solution including hardware, software, and installation for a particular application that will not change dramatically and does not require the features of a full OSI system.

- It is often easier to do special engineering on installed closed systems when small changes are required than to face the expense and problems of a new system.

- The company may plan on never allowing a multivendor environment.

- Special requirements of speed, physical environment, and application may preclude the use of OSI-compliant products.

It is interesting that while cost has been cited as one of the prime reasons for using proprietary networks rather than MAP/ TOP, a recent study by General Motors indicated that a MAP node cost GM $3000 and a proprietary node cost $2664. (2) When all factors such as porting software and training were taken into account, however, the cost of a proprietary network was 40% higher than the cost of a MAP network. These costs are for networks in an integrated company or factory environment. Small office networks could be much less expensive.

They are difficult to install and expensive, but integrated information systems are essential to corporations that hope to compete in today's marketplace. Anticipating change and planning to manage that change effectively are keys to success. Today, standards based on open communications systems like MAP/TOP provide the best path to graceful enhancement of information management while maintaining a stable base of operation.

REFERENCES

1. GM internal study.

2. GM internal study.

Section Seven: Justification

Section Seven's paper is entitled:

• *World Class Cost Management and CIM Justification: The Way of the 90s.*

Planning without positive business financial considerations is not acceptable in today's competitive environment. Nellemann maintains that long-term strategic objectives outweigh short-term demands on earnings in the global marketplace. The use of Computer-Integrated Manufacturing systems is a long term strategy that is under-used. A world class cost management system can assist the corporation in meeting the manufacturing challenges. He summarizes justification alternatives to improve a planner's ability to gain business approval for system implementation.

Presented at the CASA/SME AUTOFACT '89 Conference, October 30-November 2, 1989

World Class Cost Management and CIM Justification: The Way of the 90s

DAVID O. NELLEMANN
Andersen Consulting

INTRODUCTION

Reconciling long-term strategic objectives with short-term demands on earnings can be difficult at best and paralyzing at worst. Computer-integrated manufacturing is a long-term strategic proposition and an increasingly necessary one in this evolving global economy. Yet many managers are reluctant to aggressively pursue it. That is because traditional cost measurements overemphasize short-term quantitative factors—such as financial return—but undervalue long-term qualitative factors—like quality and flexibility—that further create marketplace advantages. The result, too often, is indecision. World class cost management is a proven, practical model that supports the realities of today's advanced manufacturing. This paper describes the elements necessary for a world class cost management system.

As our economy globalizes, companies are feeling the heat of world class competition. Particularly in manufacturing, new concepts are gaining increased acceptance, challenging the old and familiar, and accelerating the pace of change.

As managers strive to improve their competitive position, they face a serious dichotomy. On the one hand, they are pressured to rethink the way they do business—using Computer Integrated Manufacturing (CIM) as one strategic option to cut costs, increase quality, and shorten lead times.

On the other hand, managers are faced with pressure to improve profitability and return on investment. Further contributing to this dilemma are outmoded investment justification and cost management approaches.

How do managers resolve these conflicting forces and implement business strategies that offer profitable growth and long-term business survival? The basic issue is the coordination of strategic objectives with financial justification and cost management. It does not make sense to establish broad competitive strategies, then conduct narrow cost justification based on inaccurate cost information, and not have an effective performance measurement system to ensure successful implementation.

Several recent studies, in fact, show clearly that managers are postponing CIM investments because of the uncertainty of its benefits. They are unwilling to invest on "feel" alone.

The result, too often, is indecision. It forces managers into a reactive mode that drains competitive strength. In addition, there is a cost to doing nothing, which traditional financial justification analysis does not take into account.

In our experience, the best solution is to view CIM primarily as a strategic commitment, affecting all areas of the organization. As such, CIM is justified by its contribution to long-term corporate objectives, not its short-term earnings return. Unless all areas of the organization buy into the CIM program, it will become a piecemeal approach doomed to failure.

Viewing CIM as a strategy to yield significant improvements:

- Avoids the pitfall of implementing technology in isolation, or incrementally, without a vision of how best to integrate various elements.
- Sees CIM as a business strategy for the entire logistics "pipeline," in contrast to the traditional justification view of manufacturing as a series of isolated functions.
- Relates CIM to what is really important in today's intensely competitive environment—adding product value by reducing costs, improving product quality, and reducing lead time. Shrinking the pipeline by eliminating non-value-added activities lowers costs and improves marketplace responsiveness.
- Justifies CIM through the right kind of cost infor-

mation that attaches the right kind of values to key competitive factors.

- Implements CIM, once justified, through effective management reporting and performance measures consistent with the strategy.

Before CIM investments can be analyzed, key questions must be resolved:

- What is the true cost of our products?
- How much will it cost to make a product in our new plant using new technology?
- Can we make this product at a cost that will yield an acceptable return?
- What is the cost of product complexity?
- Why do overhead costs keep increasing? How can we manage them?

Obtaining accurate product costs that relate to the complete pipeline is not a new management dilemma. Measuring success is also a part of this dilemma. While many managers have moved aggressively to change their manufacturing methods, they have not changed their methods of measuring and reporting performance. As a result, improvements are allowed to migrate to their original state. Most managers have read all the articles bashing traditional cost management systems and they agree with them. But they are still looking for practical answers.

This article, therefore, offers a proven, practical alternative—World Class Cost Management that supports the realities of today's advanced manufacturing. It has been tested successfully in scores of real world situations.

WORLD CLASS COST MANAGEMENT

Since World Class Cost Management involves a combination of systems and control techniques, it is important to understand its essential elements as illustrated in Figure 1.

The basic structure of the system focuses on costs in three key areas of the business—strategic, operational, and financial. There should be only one World Class Cost Management system in the company, but it must be responsive to these three views.

- Strategic (Product Cost)—For product pricing, make/buy analysis, capital equipment investments, capacity utilization, productivity, and marketing decisions.
- Operational (Performance Measurement)—For analysis of productivity, identification of cost drivers (financial and nonfinancial), and control of value-added and non-value-added costs.
- Financial (Financial Reporting)—For valuing inventory, costing sales, and external reporting for complying with various regulations.

These three views of costs provide the underpinnings for two key information systems—management control and reporting, and cost accounting.

- Management Control and Reporting—A system providing the critical information at all management levels to ensure responsive and coordinated direction for the company.
- Cost Accounting—Represents the support system and changes that need to be made to focus on product lines and continuous improvement, while still meeting the requirements of financial reporting.

Since there tends to be a good deal of confusion in current literature as to the basic elements and techniques embodied in any cost management system, much less a World Class Cost Management system, it is useful to look at those areas affecting product costing and performance measurement in a little more depth.

Product Costing

Developing and maintaining accurate product costs is vital in making certain strategic choices such as: 1) product pricing, 2) make/buy, 3) investment justifications, and 4) marketing strategies.

In response to this need, a technique called Activity Based Costing has proven useful and is gaining a lot of attention. Essentially, it is a better way to apply overhead, one that addresses costs across the full company "pipeline"— encompassing preproduction, production, and postproduction activities.

Traditional systems of allocating overhead often provide a distorted picture of product costs, frequently resulting in erroneous decisions. Activity Based Costing, on the other hand, deals with the traditional problem of applying larger and larger overhead percentages to smaller and smaller components of cost. It is used to determine all of the major costs associated with a specific product as it moves through the logistics pipeline.

To illustrate the technique, consider the case of a manufacturer of machined parts that we worked with recently on a classic problem: management questioned the profitability of serving a key customer, and whether in the long run they wanted to be in this product line. They felt existing costing methods were inadequate to make confident judgments.

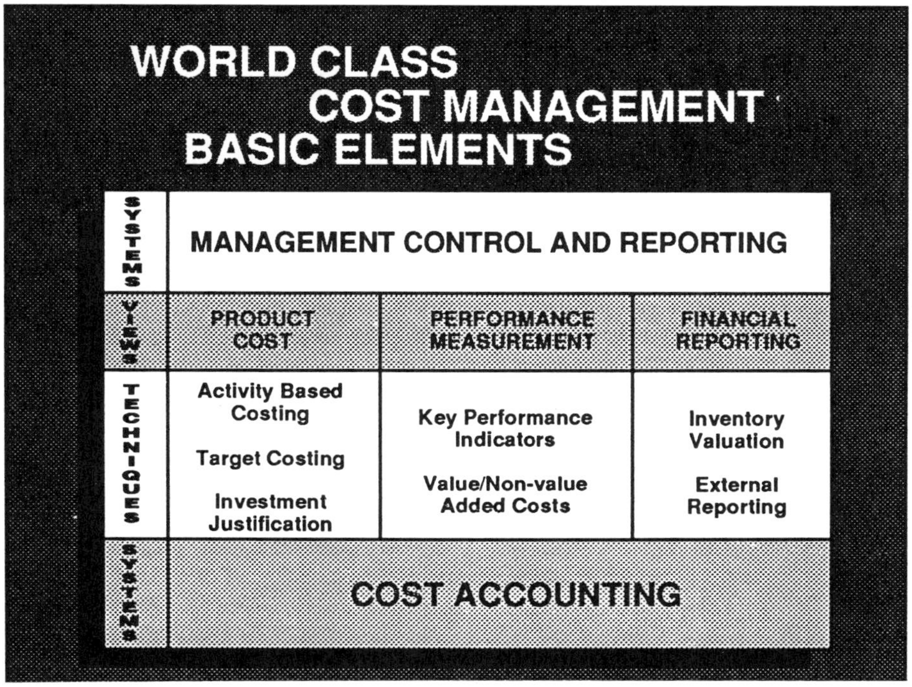

WORLD CLASS
COST MANAGEMENT
BASIC ELEMENTS

SYSTEMS	MANAGEMENT CONTROL AND REPORTING		
VIEWS	PRODUCT COST	PERFORMANCE MEASUREMENT	FINANCIAL REPORTING
TECHNIQUES	Activity Based Costing Target Costing Investment Justification	Key Performance Indicators Value/Non-value Added Costs	Inventory Valuation External Reporting
SYSTEMS	COST ACCOUNTING		

Figure 1.

Through Activity Based Costing techniques, we developed a product line profitability model. It showed, among other things, that certain products thought to be profitable were not, and vice versa. Specifically, it highlighted the unusually high fixed overhead, principally in equipment, consumed by the product line under review, and the impact on profitability of lower than expected order levels over the long run.

As illustrated in Figure 2, the traditional system took a large pool of overhead costs and allocated them on a percentage basis to direct material and labor charges, producing a gross profit of 23 percent. Under Activity Based Costing, however, overhead costs were allocated directly, based on an analysis of the real cost drivers. As a result, gross profit turned out to be only 11 percent.

With this kind of information on a key product line, the company could make significant changes affecting the future, including renegotiation with the key customer.

PRODUCT: CONNECTOR (1000 UNITS)

	BEFORE	REVISED
Sales price	3122	3122
Direct Materials, Outside Purchases & Labor	1054	1054
	2068	2068
Manufacturing Overhead		
10% of Direct Materials	40	0
15% of Outside Purchases	16	0
234% of Direct Labor	1281	0
Supervision/Management/Inspection	0	434
Engineering/Production Planning	0	249
Equipment/Facilities	0	531
Purchasing & Distribution	0	272
Supplies/Maintenance	0	248
Total Manufactured Overhead	1337	1734
Gross Profit	731	334
%	23%	11%

Figure 2.

Investment Management

Once accurate product costs, whether historical or target, are developed, a number of other factors must enter into the investment justification process.

Because plants and equipment must be replaced, the plan for introducing and integrating new technology must center on the rate and priority of investments. This is where financial analysis and World Class Cost Management systems play an important role.

Integrating business strategy, economic justification, and performance measurement historically has been a major management failing. Too often we see broad strategies (which may include CIM) to achieve competitive advantage, but justification, as well as managers' performance, is measured along very narrow lines. Examples frequently encountered in industry include:

- Bias toward incremental investments. Typical financial analysis is based on piecemeal introduction of new equipment, with paybacks reflecting the anticipated incremental benefits of each machine. The long-range benefits of technological integration are ignored.
- Inability to simulate and analyze alternative business scenarios. For many companies, the only vision of the future is based on the facts of the past. Managers need tools to evaluate alternatives, given different assumptions about the future.
- Failure to quantify important benefits and assess opportunity costs. Savings in quality, flexibility, customer service, and space are often ignored. While not as easily identified as labor and inventory reductions, these costs are no less important. In addition, managers need to know the opportunity cost of doing nothing. What, for example, is the impact of maintaining the status quo when competitors are investing?
- Excessively high hurdle rates not tied to product and market strategies. Should investments in emerging products with high potential be held to the same hurdle rates as investments in mature products with limited market potential?

Traditional cost justification procedures contain a single hurdle rate for all investment decisions, irrespective of product lines and their market positions. Yet manufacturers that work toward product and process simplification focus operations on a product-line basis. Figure 3 sets out relevant guide lines used by one company.

	-------- Product --------		
	Strategic	New	Cash
Market Potential	High	Medium	Low
Life Cycle Stage	Growing	Emerging	Maturing
Investment Hurdle Rate	8%	12%	16%

Figure 3.

THE "FRESH START" APPROACH

A ''fresh start'' justification analysis compares a company's competitive position assuming two scenarios:

- A total CIM environment.
- The status quo (only modest technology upgrades).

In both cases, the company starts from scratch by:

- Acquiring new plants and equipment.
- Hiring new employees with appropriate skills.
- Investing in necessary hardware and software.

Since CIM improves flexibility, speed, and quality, a situation of stable or increasing market share is assumed for the 100 percent CIM scenario. The status quo scenario assumes, however, that competitors invest in technology and thus the company faces a declining market share.

These two scenarios illustrate the full impact and opportunity costs of the two baseline strategies. The premise is that, if cash flows support the use of a CIM environment, the conversion will make sense—even if traditional investment justification suggests otherwise. As a result, decisions can focus on strategy execution.

The product represented a strategic opportunity to invest in a related product line with solid growth potential. While some existing fully depreciated equipment was available for the manufacture of the new product, the decision was made to avoid taking a short-term approach based solely on traditional financial justification techniques. Instead, an attempt was made to assess the longer range strategic implications of the investment decision.

Assumptions

The assumptions were as follows:

- A CIM strategy would increase market share gradually from 20 percent to 24 percent over the ten-year life cycle, and unit prices would decline from $10.00 to $8.50.
- A status quo strategy would lose market share because CIM investments would give competitors greater flexibility and higher product values. It was assumed that market share would, therefore, decrease from 20 percent to 15 percent and that unit prices would fall from $10 to $8.

While material costs would be the same for both alternatives, the CIM environment would:

- Cut direct labor costs 30 percent.

- Reduce cost of quality by 60 percent.
- Improve inventory turns from 3.5 to 10, lowering inventory carrying costs.
- Raise occupancy costs 33 percent, reflecting greater power and environmental needs.
- Raise engineering and technical support 50 percent.
- Cut plant management and supervisory costs 25 percent, due to lower head counts and improved information networks.

Results

A cash flow model, Figure 4, shows that, while payback occurs a year later under the CIM alternative, total cumulative cash flows are three times those of the status quo alternative by year 10. Moreover, if cash flows are discounted, cumulative cash flows never exceed zero for the status quo alternative.

Figure 5 shows the dramatic long-range impact of the two strategies. Under the status quo alternative, contribution margin falls to zero in year 10 while hovering at 20 percent under the CIM alternative. Although unit volume is largely responsible for the difference, the CIM alternative enjoys a 17 percent advantage in non-value-added costs.

In both cases, margins decline over the product life cycle due to competitive pressures. The market, however, remains viable and profitable in year 10 (and beyond) under a CIM strategy; it is far less attractive if the status quo strategy is followed.

Initial Investments

The initial investment to launch production for the status quo alternative was determined to be $19 million, compared with $28 million for CIM.

Under CIM, because of better space utilization, a smaller plant would be needed initially. The total costs for equipment and tooling, computer hardware and software, and implementation, however, were assumed to be nearly twice as much as the corresponding costs for the status quo scenario. Further, because of market growth and increasing market share in the CIM scenario, increased capacity that would cost another $13 million would be incurred in year three. No additional capital investments were assumed for the status quo alternative.

Long-Term View

Figure 6 shows how deceiving a short-range ROI approach to justification can be. In the first four years, the ROI under the status quo scenario is higher than the ROI under the CIM alternative. As status quo market share continues to drop, however, so does ROI. By year six, if ROI is used as the yardstick, it becomes clear that the CIM scenario is the clear winner.

The fresh start approach forces managers to step back and consider all aspects of an investment equation, including both revenue and cost components. By comparing alternatives, the fresh start approach allows managers to evaluate the long-term implications of major strategic investments in technology.

PERFORMANCE MEASUREMENT

The real challenge, however, is implementing the

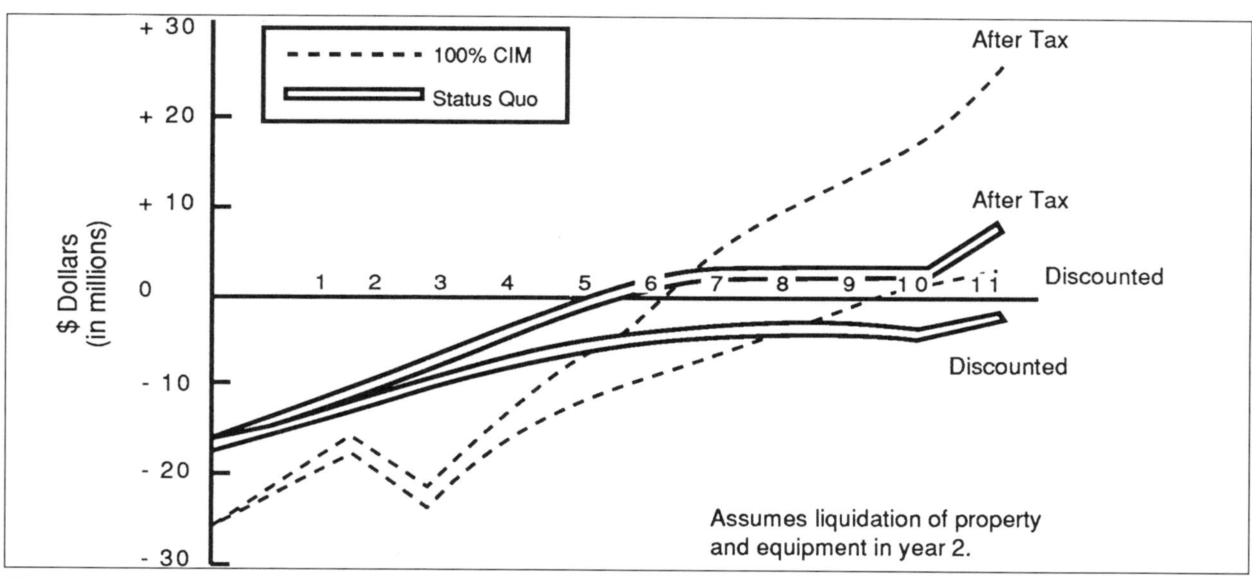

Figure 4.

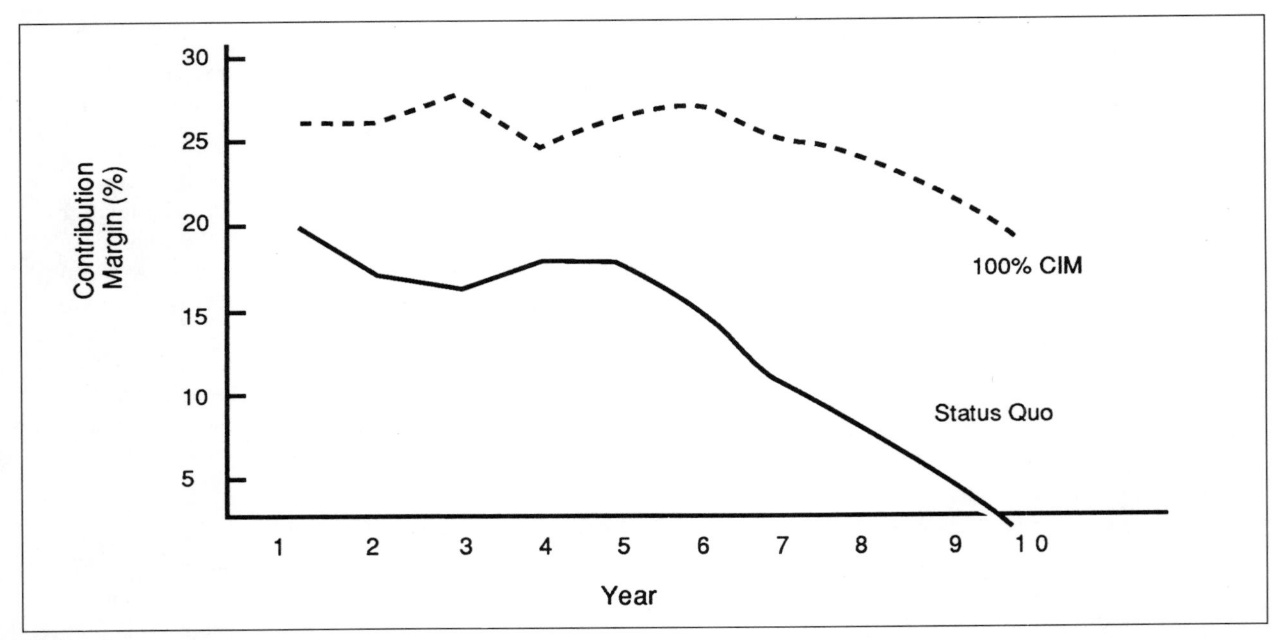

Figure 5.

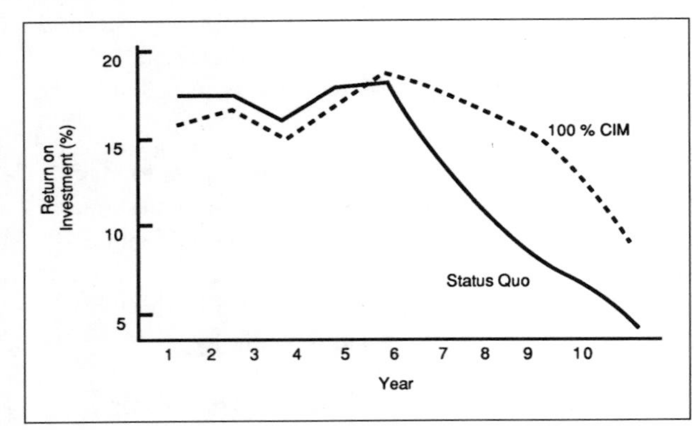

Figure 6.

CIM strategy and realizing the benefits. Success depends on skilled and committed managers who keep their eyes on the objectives while managing day-to-day activities.

New standards of performance must be the rule in today's manufacturing environment. Both financial and nonfinancial strategic objectives should be defined to support daily business decisions and to measure performance throughout the organization. The linkage between setting company objectives as expressed by key success factors and the development of key performance indicators (KPIs) for all operations and functions is illustrated in Figure 7.

Once management determines how critical it is to achieve certain market and financial objectives, business strategies can be formulated.

Management at all levels has to determine what measures it will use to achieve high quality, implement quick design changes, improve delivery lead time, or reduce cost to meet the company's overall objectives. Once these factors have been determined, they can be "translated" into specific measures of each business function in the logistics pipeline.

Case in point: A manufacturer of carbide tools and tooling systems recently asked us to help develop an information system to support its new CIM strategy of providing greater product and service diversity globally.

Our approach was to improve operational reporting at all management levels by translating the strategics into the appropriate key performance indicators by focusing on cost creating activities. Their responsibilities for the KPls were assigned at levels where they could be controlled.

In this way, employees focus on the real cost generators, instead of those that are merely symptomatic of the problem, e.g., labor and overhead variances. In effect, the system encourages management to work toward the common objectives set out in their CIM justification process.

In summary, these are the characteristics of key performance indicators:

- Are designed to encourage the achievement of

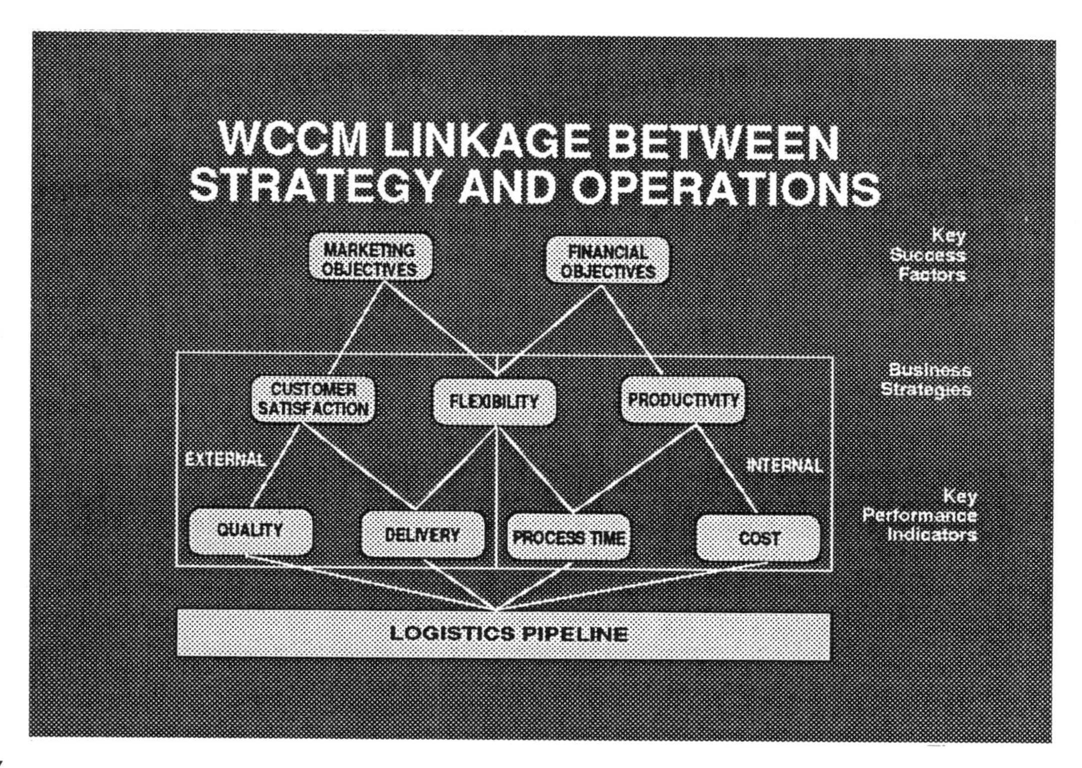

Figure 7.

strategic objectives by providing timely focused feedback.

- Apply to all levels of management.
- Are not complex and help to create a common language between operations, finance, and other functions.

Figures 8 and 9 illustrate how various key performance indicators and reporting characteristics might differ at the division and manufacturing cell levels.

GETTING STARTED

Melding strategies, financial justification, and performance measurement together is essential in a World Class Cost Management system.

Moving to World Class Cost Management is a significant undertaking and requires a well planned approach. It is vital to get started right. Trying to do too much at once can seriously disrupt an organization. It is better to use a step-by-step approach, gaining management buy-in at all levels. Managers must recognize that a major cultural change is needed to effect meaningful improvements, because new approaches will be required and people will be measured differently.

A good place to start is to develop methods to revise product costs, revise financial justification procedures, and adopt the fresh start approach for CIM justification. This could be followed by the design and implementation of a new management control and reporting system. It may also make sense to start with a pilot project, based on a specific product line or single plant or division.

PUTTING THE PIECES TOGETHER

The reluctance of many manufacturers to make major investments in CIM stems from several factors:

- A view of CIM as a manufacturing program instead of a company-wide strategy.
- Inappropriate techniques for justifying and measuring the success of the investments.

The result has been low-risk, low-growth decisions based on short-term financial pressures. Status quo thinking has sapped the industrial competitive strength of many companies to the point that their very survival is threatened.

Progressive companies, on the other hand, are looking at CIM differently. They see it as a vital link in a company's logistics pipeline, integral to the company's overall strategy. Traditional justification approaches that encourage incremental, short-range investments must be scrapped. The "fresh start" approach assesses

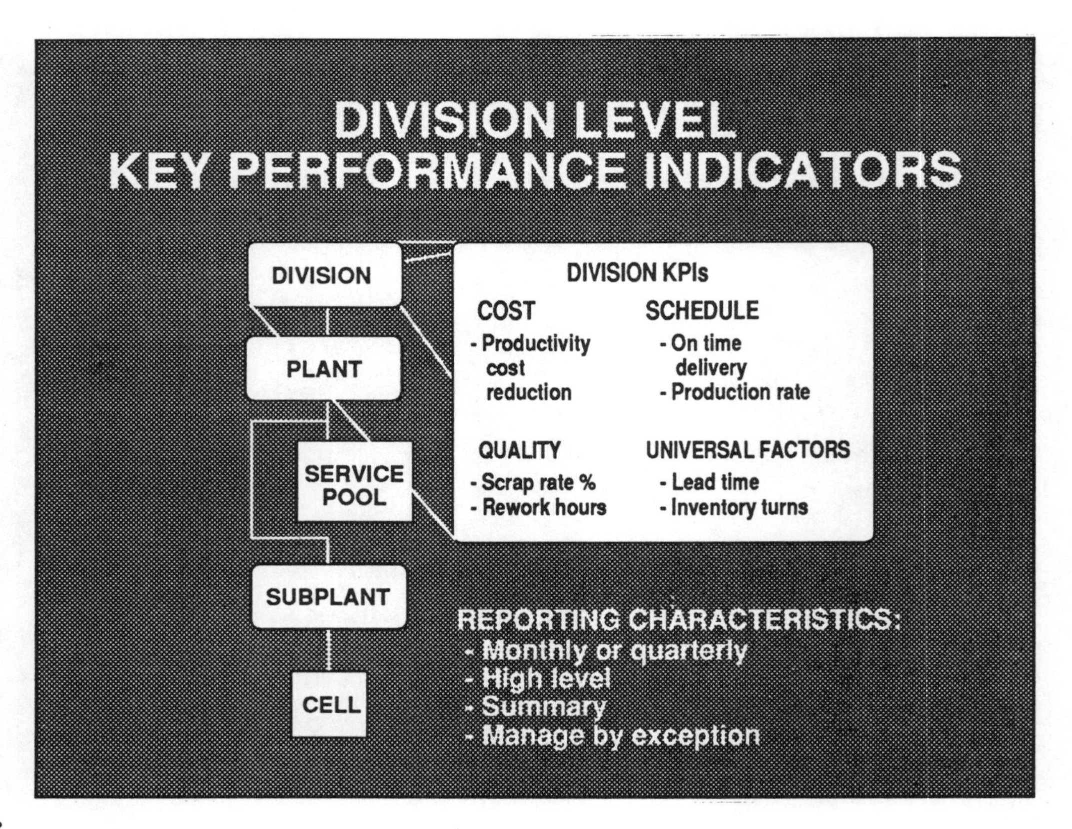

Figure 8.

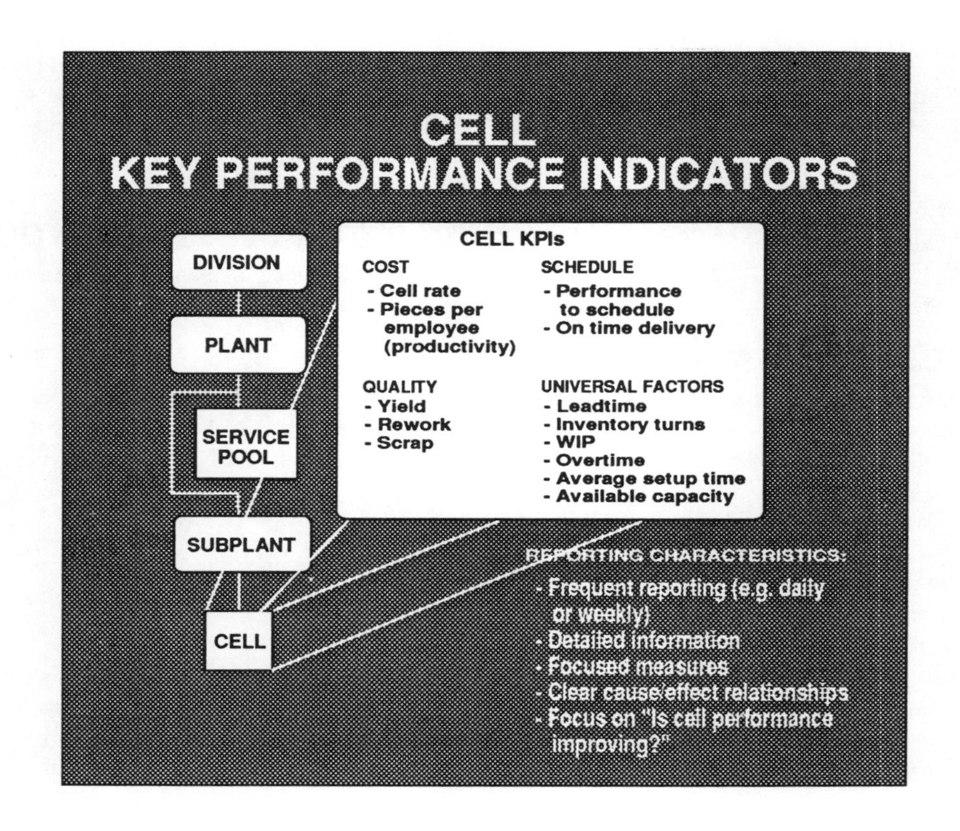

Figure 9.

the opportunity costs of not investing and identifies all costs and benefits.

Once justified, the success of a CIM strategy then depends on effective and ongoing performance measures that provide incentives consistent with the strategy, as illustrated in Figure 10.

World Class Cost Management, incorporating appropriate CIM justification procedures, represents a framework for developing the information systems necessary to manage the organization in a manner that is responsive to today's competitive world. In summary, the benefits of World Class Cost Management are threefold:

- It facilitates the identification and implementation of company strategies.
- It provides more accurate and easy-to-use information.
- It helps reduce the costs associated with traditional cost management systems.

World Class Cost Management is becoming the competitive tool of the 90s. Without it, no organization can become a World Class competitor.

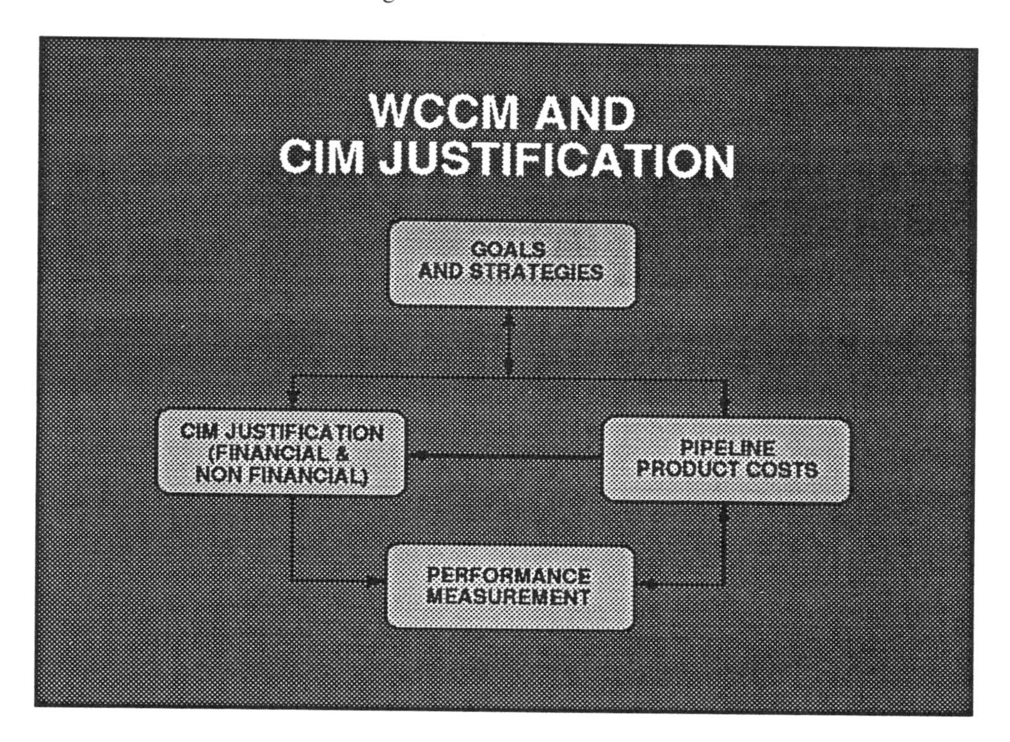

Figure 10.

Section Eight:
Reference Bibliography

Reference Bibliography

A good understanding of EIX cannot be achieved from a single source. This reference bibliography is provided as recommended information sources that will enhance the Manager, Manufacturing Engineer, and/or Systems Planner's knowledge of EIX. The reference bibliography contains the following subsections:

General EIX Periodical References
EDI Periodical References
PDES Periodical References
IGES Periodical References
CALS Periodical References
EDI Book References
PDES Book References
IGES Book References
CALS Book References
Government Publications References

GENERAL EIX PERIODICAL REFERENCES

Anderson, Sandy. "How Information Engineering Helped CILCO Beat the Clock." *Chief Information Officer Journal*, Vol. 4, No. 3, pp. 61-64, Winter 1992.

Anonymous. "Coopers and Lybrand Creates BBS to Link Business Incubators." *Communications News*, Vol. 29, No. 3, p. 30, March 1992.

Appleton, Daniel S. "Understanding Today's Enterprises." *Manufacturing Systems*. Vol. 10, No. 1, pp. 52-57, January 1992.

Balk, Walter; Olshfski, Dorothy; Epstein, Paul; and Holzer, Marc. "Perspectives on Productivity." *Public Productivity and Management Review*, Vol. 15, No. 2, pp. 265-279, Winter 1991.

Bird, Jane. "Managing Information Technology: Let Your Fingers Do the Walking." *Management Today*, pp. 97-104, November 1991.

Blum, Daniel. "Crossing the Enterprise with LAN E-mail Messages." *Network World*, Vol. 9, No. 9, pp. 1, 32-37, 39-40, 47-51, March 2, 1992.

Cashin, Jerry. "Blueprints for Openness: The Ink Is Still Drying." *Software Magazine*, Vol. 11, No. 6, pp. 92-102, May 1991.

Chaykowski, Richard P. and Slotsve, George A. "The Impact of Plant Modernization on Organizational Work Practices." *Industrial Relations*, Vol. 31, No. 2, pp. 309-329, Spring 1992.

Currid, Cheryl. "IS Must Include Extra-Enterprise Information Needs in Plan." *InfoWorld*, Vol. 14, No. 20, pp. 60, May 18, 1992.

Dileva, A.; Vernadat, F.; and Bizier, D. "From Enterprise Analysis to Conceptual Database Design in Production Systems." *Engineering Costs and Production Economics*, Vol. 16, No. 3, pp. 171-182, June 1989.

Fuld, Leonard M. "Achieving Total Quality Through Intelligence." *Long Range Planning*, Vol. 25, No. 1, pp. 109-115, February 1992.

Gallupe, Brent and Fox, George. "Facilitated Electronic Meetings: Higher Quality, Less Time," *CMA Magazine*, Vol. 66, No. 3, pp. 29-32, April 1992.

Ghoshal, Sumantra and Butler, Charlotte. "The Kao Corporation: A Case Study." *European Management Journal*, Vol. 10, No. 2, pp. 179-192, June 1992.

Grantham, Tim. "Big Companies Persist in Push for Open Systems." *Computerworld*, Vol. 26, No. 11, pp. 57-58, March 16, 1992.

Heide, Jan B. and Miner, Anne S. "The Shadow of the Future: Effects of Anticipated Interaction and Frequency of Contact on Buyer-Seller Cooperation." *Academy of Management Journal*, Vol. 35, No. 2, pp. 265-291, June 1992.

Hein, K. Peter. "DevelopMate: A New Paradigm for Information System Enabling." *IBM Systems Journal*, Vol. 29, No. 2, pp. 250-264, 1990.

Hsu, Cheng; Bouziane, Mohamed; Rattner, Laurie; and Yee, Lester. "Information Resources Management in Heterogeneous Distributed Environments." *IEEE Transactions on Software Engineering*, Vol. 17, No. 6, pp. 604-625, June 1991.

Janson, Philippe; Molva, Refik; and Zatti, Stefano. "Architectural Directions for Opening IBM Networks: The Case of OSI." *IBM Systems Journal,* Vol. 31, No. 2, pp. 313-335, 1992.

Krouse, John; Mills, Robert; Beckert, Beverly; Carrabine, Laura; and Berardinis, Lawrence. "Putting Integration to Work." *Industry Week,* Vol. 240, No. 14, pp. CC34-CC39, July 15, 1991.

Johnson, Maryfran. "IS Eyes Business Intelligence." *Computerworld,* Vol. 25, No. 42, pp. 31, October 21, 1991.

Lambert, Peter. "Imaging: The Merger of Computer and TV." *Broadcasting,*Vol. 122, No. 5, pp. 41-42, January 27, 1992.

Lehman, Tom. "Information Networks and Innovative Approaches." *Public Productivity and Management Review,* Vol. 15, No. 2, pp. 245-251, Winter 1991.

MacLean, Lisa M. "What to Consider Before Leaping into a LAN." *Office Systems,* Vol. 8, No. 12, pp. 12-21, December 1991.

Manley, John H. "Paraprogramming Manufacturing Information Systems." *Computers and Industrial Engineering,* Vol. 21, No. 1-4, pp. 229-233, 1991.

Mantelman, Lee. "Workflow: Improving the Flow of Corporate Data." *InfoWorld,* Vol. 14, No. 26, pp. 46-47, June 29, 1992.

Mazda, Fraidoon. "Standardizing On Standards." *Telecommunications (International Edition)* Vol. 26, No. 6, pp. 40-50, June 1992.

Metcalf, Lynn E.; Frear, Carl R.; and Krishnan, R. "Buyer-Seller Relationships: An Application of the IMP Interaction Model." *European Journal of Marketing,* Vol. 26, No. 2, pp. 27-46, 1992.

Oliff, Michael D. and Marchand, Donald A. "Strategic Information Management in Global Manufacturing." *European Management Journal,* Vol. 9, No. 4, pp. 361-372, December 1991.

Orr, Joel N. "Join the Information Economy." *CAE,* Vol. 11, No. 4, pp. 84, April 1992.

Parker, M. M. "Enterprise Information Analysis: Cost-Benefit Analysis and the Data-Managed System." *IBM Systems Journal,* Vol. 21, No. 1, pp. 108-123, 1982.

Pilling, Bruce K. and Zhang, Li. "Cooperative Exchange: Rewards and Risks." *International Journal of Purchasing and Materials Management,* Vol. 28, No. 2, pp. 2-9, Spring 1992.

Rapaport, Matthew. "Groupware vs. CCS: Comparing Benefits and Functionality." *Telecommunications,* Vol. 25, No. 11, (North American Edition) pp. 37-40, November 1991.

Richardson, Gary L.; Jackson, Brad M.; and Dickson, Gary W. "A Principles-Based Enterprise Architecture: Lessons from Texaco and Star Enterprise." *MIS Quarterly,* Vol. 14, No. 4, pp. 385-403, December 1990.

Romei, Lura K. "Multimedia: Delivering Added Value Today." *Modern Office Technology,* Vol. 36, No. 10, pp. 43-46, October 1991.

Ryan, Hugh W. "Experimenting with the Envelope." *Journal of Information Systems Management,* Vol. 5, No. 2, pp. 72-79, Spring 1988.

Sheridan, John H. "The CIM Evolution." *Industry Week,* Vol. 241, No. 8, pp. 29-51, April 20, 1992.

Sloan, Ken. "All Your Assets On-Line." *Computerworld,* Vol. 22, No. 28, pp. 55-58, July 11, 1988.

Snyder, James D. "Integration Becoming the Key Ingredient in Evolution of Environmental Software." *Environment Today,* Vol. 3, No. 2, pp. 29, 32, March 1992.

Tavora, Carlos. "Flexible Systems: A Configuration Control Challenge." *Manufacturing Systems,* Vol. 7, No. 11, pp. 34-40, November 1989.

Tomkinson, Donald. "Getting MEs and EEs to Work in Harmony." *Machine Design,* Vol. 64, No. 2, pp. 60-64, January 23, 1992.

Weimer, George; Knill, Bernie; Beckert, Beverly; and Teresko, John. "Integrated Manufacturing: Systems Integrators Put It All Together." *Industry Week,* Vol. 240, No. 22, pp. IM3-IM15, November 18, 1991.

Zikiye, Anthony A. and Zikiye, Rebecca A.

"Satisfaction Gaps: New Realities in Managing Automation." *Management Decision*, Vol. 30, No. 2, pp. 40-45, 1992.

EDI PERIODICAL REFERENCES

Anonymous. "EDI Gets Boost with New Products, Vendor Alliances." *Network World*, Vol. 9, No. 21, pp. 23-24, May 25, 1992.

Koester, Larry M. "NAFTA: World Class Opportunity." *Transportation and Distribution*, Vol. 33, No. 2, pp. 42-44, February 1992.

Allen, Brandt R. and Boynton, Andrew C. "Information Architecture: In Search of Efficient Flexibility." *MIS Quarterly*, Vol. 15, No. 4, pp. 435-445, December 1991.

Cashin, Jerry."APP Offers Guidelines for Federal Purchasing." *Software Magazine*, Vol. 11, No. 10, pp. 84-95, August 1991.

Cahn, David M. "What Benefits Will Global EDI Offer?" *Transportation and Distribution*, Vol. 33, No. 6, pp. 63-64, June 1992.

Cerf, Vinton G. "Prospects for Electronic Data Interchange." *Telecommunications*, Vol. 25, No. 1, (North American Edition) pp. 57-60, January 1991.

Etheridge, James. "EDI in Europe." *Datamation*, Vol. 34, No. 19 (International Edition), pp. 44-15, 44-18, October 1, 1988.

Edwards, Robert D.; Saxer, Craig S.; and Edwards, Thomas J. "EDI Payment Development: The Dollars and Data Issue." *Journal of Cash Management*, Vol. 10, No. 1, pp. 17-22, January/February 1990.

Forger, Gary. "Making Bar Coded Labels and EDI Pay Off." *Modern Materials Handling*, Vol. 46, No. 7, pp. 57-59, June 1991.

Freiwirth, Raphael R. "Standards: Shopping for X.400." *Networking Management*, Vol. 10, No. 9, pp. 48-51, August 1992.

Frye, Colleen. "EDI Beginning to Stretch National, Business Bounds." *Software Magazine*, Vol. 12, No. 6, pp. 88-91, May 1992.

Knill, Bernie. "Continous Flow Manufacturing: Driver of the Automatic Factory." *Material Handling Engineering*, Vol. 47, No. 5, pp. 54-57, May 1992.

Laplante, Alice. "Troubled Automakers Rethink IS." *Computerworld*, Vol. 26, No. 3, pp. 67-68, January 20, 1992.

Mandell, Mel. "EDI Speeds Caterpillar's Global March." *Computerworld*, Vol. 25, No. 32, pp. 58, August 12, 1991.

Miley, Mike. "X.400 Pushes the Envelope." *InfoWorld*, Vol. 14, No. 29, pp. 42-43, July 20, 1992.

Puttre, Michael. "Electronic Data Exchange Draws Design Teams Closer Together." *Mechanical Engineering*, Vol. 113, No. 5, pp. 56-59, May 1991.

Ross, Julie Ritzer. "Developments Cut Costs, Speed Shipments." *Global Trade*, Vol. 111, No. 11, pp. 14-18, November 1991.

Tallyen, Jeffrey J. "IRS Record Retention Rules and the Computer Age." *Tax Adviser*, Vol. 23, No. 7, pp. 441-442, July 1992.

Teschler, Lee; Dreyer, Jerome L.; Van Wolvelaere, Cory; Lavery, Hank; and Tannenwald, Peter. "Presidential Report on EDI: EDI Today Focusing on Steps for Success and Competitive Advantage." *Industry Week*, Vol. 240, No. 21, pp. EDI1-EDI18, November 4, 1991.

Wayland, Fred. "EDIFACT or ANSI: Which Road Leads to EDI Goal?" *Corporate Cashflow*, Vol. 13, No. 8, pp. 37-40, July 1992.

Weintraub, Emanuel. "EDI Delivers Quick Response Distribution." *Bobbin*, Vol. 30, No. 3, pp. 22, 24, November 1988.

Witt, Clyde E. "K-mart's New Distribution Center: Quick Response in Action." *Material Handling Engineering*, Vol. 47, No. 3, pp. 46-52, March 1992

PDES PERIODICAL REFERENCES

Anonymous. "CAD/CAM Planning: Communicating Globally." *CAE*, Vol. 11, No. 7, pp. CC26-CC35, July 1992.

Dale, Chris. "Improving the CAD-FEM Connection." *CAE*, Vol. 11, No. 4, pp. 76-77, April 1992.

Krouse, John; Mills, Robert; Beckert, Beverly;

Carrabine, Laura; and Berardinis, Lawrence. "Exploiting Advanced Software." *Industry Week*, Vol. 240, No. 14, pp. CC10-CC14, July 15, 1991.

Liker, Jeffrey K.; Fleischer, Mitchell; and Arnsdorf, David. "Fulfilling the Promises of CAD." *Sloan Management Review*, Vol. 33, No. 3, pp. 74-86, Spring 1992.

Loeffelholz, Suzanne. "CAD/CAM Comes of Age." *Financial World*, Vol. 157, No. 22, pp. 38-40, October 18, 1988.

Merrick, Lew. "A Better Way to Send Engineering Drawings." *Machine Design*, Vol. 65, No. 13, pp. 44, 46, June 25, 1992.

Ohr, Stephan. "Speeding Process Planning for the Navy." *Manufacturing Systems*, Vol. 8, No. 8, pp. 26-30, August 1990.

Semakula, Mukasa E. and Satsangi, Ajay. "Application of PDES to CAD/CAPP Integration." *Computers and Industrial Engineering*, Vol. 17, No. 1-4, pp. 234-239, 1989.

Semakula, Mukasa E. and Satsangi, Ajay. "Application of PDES to CAD/CAPP Integration. "*Computers and Industrial Engineering*, Vol. 18, No. 4, pp. 435-444, 1990.

Skibinski, John. "Automated Information-Sharing Cuts Time-to-Market." *Manufacturing Systems*, Vol. 10, No. 5, pp. 60-64, May 1992.

Warthen, Barbara. "PDES Shapes Data Exchange Technology." *CAE*, Vol. 9, No. 2, pp. 68-72, February 1990.

IGES PERIODICAL REFERENCES

Anonymous. "CAD/CAM Planning: Communicating Globally." *CAE*, Vol. 11, No. 7, pp. CC26-CC35, July 1992.

Anonymous. "CAD/CAM Planning: Working in Concert." *CAE*, Vol. 11, No. 7, pp. CC14-CC22, July 1992.

Beckert, Beverly A. "Fixing Snags in File Exchange." *CAE*, Vol. 11, No. 7, p. 76, July 1992.

Beckert, Beverly A. "Molds in the Making." *CAE*, Vol. 11, No. 3, pp. 62-68, March 1992.

Dale, Chris. "Improving the CAD-FEM Connection." *CAE*, Vol. 11, No. 4, pp. 76-77, April 1992.

Eckerson, Wayne. "Paperfree Factory Gets Patriot Flying." *Network World*, Vol. 8, No. 52/Vol. 9, No. 1, pp. 1, 39, December 30, 1991/January 6, 1992.

Eshun, Thomas P.; Chen, Chin-Sheng; Owusu-Ofori, Samuel P.; and Sarin, Sanjiv. "Data Integrity in an IGES Description of Turned Part Geometry." *Computers and Industrial Engineering*, Vol. 21, No. 1-4, pp. 459-463, 1991.

Halliday, Caroline. "Product Comparison: High-End CAD." *InfoWorld*, Vol. 13, No. 16, pp. 55-75, April 22, 1991.

Hemmelgarn, Don; Mattei, David; and Carl, Edward. "The New and Improved IGES." *CAE*, Vol. 10, No. 7, pp. 82-85, July 1991.

Khol, Ronald. "Ansys Builds Strong Links to Design and Manufacturing." *Machine Design*, Vol. 64, No. 16, pp. 14, 16, August 6, 1992.

Kramer, Denise. "MicroCAD/CAM Moves Up." *Manufacturing Systems*, Vol. 10, No. 5, pp. 48-51, May 1992.

Kwok, Wai-Lun and Eagle, Paul J. "Reverse Engineering: Extracting CAD Data from Existing Parts." *Mechanical Engineering*, Vol. 113, No. 3, pp. 52-55, March 1991.

Liker, Jeffrey K.; Fleischer, Mitchell; and Arnsdorf, David. "Fulfilling the Promises of CAD." *Sloan Management Review*, Vol. 33, No. 3, pp. 74-86, Spring 1992.

Madurai, Srinivasakumar S. and Lin, Li. "Rule-Based Automatic Part Feature Extraction and Recognition from CAD Data." *Computers and Industrial Engineering*, Vol. 22, No. 1, pp. 49-62, January 1992.

Merrick, Lew. "A Better Way to Send Engineering Drawings." *Machine Design*, Vol. 65, No. 13, pp. 44, 46, June 25, 1992.

Mills, Robert. "Finite-Element Modelers: Friendly Faces for FEA." *CAE*, Vol. 11, No. 3, pp. 36-54, March 1992.

Puttre, Michael. "Electronic Data Exchange Draws

Design Teams Closer Together." *Mechanical Engineering,* Vol. 113, No. 5, pp. 56-59, May 1991.

Skibinski, John. "Automated Information-Sharing Cuts Time-to-Market." *Manufacturing Systems*, Vol. 10, No. 5, pp. 60-64, May 1992.

Wohlers, Terry T. "CAD Meets Rapid Prototyping." *CAE*, Vol. 11, No. 4, pp. 66-74, April 1992.

CALS PERIODICAL REFERENCES

Anonymous. "Aerospace Manufacturer Launches New Information Strategy." *Industrial Engineering*, Vol. 24, No. 8, pp. 20, 22, August 1992.

Anonymous. "The War Against Paper." *Manufacturing Engineering*, Vol. 104, No. 5, pp. 20-24, May 1990.

Beckert, Beverly A. "Technical Publishing: Software Delivers on Documentation." *CAE*, Vol. 9, No. 7, pp. 100-102, July 1990.

Beazley, William G. "CALS and Its Impact." *Inform*, Vol. 4, No. 9, pp. 12-14, 16-18, October 1990.

Cashin, Jerry. "Business Transactions Take Electronic Route." *Software Magazine*, Vol. 11, No. 15, pp. 81-84, 86, December 1991.

Davis, Dwight B. "DOD Wants to Throw Tons of Paperwork Overboard." *Electronic Business*, Vol. 16, No. 3, pp. 15-16, February 5, 1990.

Gnerre, William. "EDMICS —The Foundation for Future CALS Initiatives." *Document Image Automation*, Vol. 11, No. 6, pp. 327-332, November/December 1991.

Greene, Alice H. "CALS—Understanding the Initiative. Production and Inventory." *Management Review*, Vol. 10, No. 3, pp. 34-35, March 1990.

Gribb, Thomas W. "CALS Program Offers Tech Doc Benefits." *Electronic Publishing and Printing*, Vol. 5, No. 4, pp. 42-47, May 1990.

Karsh, Arlene. "We the People: Standards Give Government More Bang for Our Bucks." *Electronic Publishing and Printing*, Vol. 4, No. 5, pp. 36-48, June/July 1989.

Kaebnick, Gregory E. "Engineering Update: An Affair with EIM —But Will They Respect It Tomorrow?" *Inform*, Vol. 3, No. 9, pp. 18-20, September 1989.

Kaebnick, Gregory E. "EDMICS Goes to Work." *Inform*, Vol. 4, No. 9, pp. 15-16, October 1990.

Korzeniowski, Paul. "Paperless Office Deluged with Paper." *Software Magazine*, Vol. 10, No. 2, pp. 66-69, February 1990.

Laplante, Alice. "Troubled Automakers Rethink IS." *Computerworld*, Vol. 26, No. 3, pp. 67-68, January 20, 1992.

Mooney, Mike. "EDMICS—Setting an Imaging Standard for the Department of Defense." *Inform*, Vol. 5, No. 10, pp. 12-15, 46-48, November/December 1991.

Murray, Helen Gagianas. "Meeting CALS Requirements." *Inform*, Vol. 3, No. 11,12, p. 27, November/December 1989.

Naughton, Michael. "CALS, VANS, and EDI." *Telecommunications (International Edition)*, Vol. 23, No. 6, pp. 67-70, June 1989.

Nordwall, Bruce D. "Digital Data System Expected to Benefit Defense and Industry." *Aviation Week and Space Technology*, Vol. 132, No. 6, pp. 66-70, February 5, 1990.

Pechersky, Paul N. "Easing Paper Flow Could Help Slash DOD Budget." *Computerworld*, Vol. 23, No. 33, pp. 67-72, August 14, 1989.

Pennington, Mike. "The Federal Forefront of Electronic Imaging." *Inform*, Vol. 3, No. 11,12, pp. 26, 28-30, November/December 1989.

Puttre, Michael. "The CALS Connection." *Mechanical Engineering*, Vol. 113, No. 8, pp. 56-59, August 1991.

Rupp, Dale. "CALS —An Initiative for Creating a Digital Standard Format." *IMC Journal*, Vol. 26, No. 4, pp. 6-8, July/August 1990.

Rupp, Dale O. "The CALS Initiative: Creating a Standard Digital Format." *Inform*, Vol. 4, No. 1, pp. 31-33, January 1990.

Ryan, Hugh W. "Open Systems: A Perspective on Change." *Journal of Information Systems Management*, Vol. 8, No. 2, pp. 62-66, Spring 1991.

Stover, Richard N. "Data Exchange: Standards for Documentation Management." *CAE*, Vol. 8, No. 8, pp. 52-55, August 1989.

EDI BOOK REFERENCES

Baker, Richard H. *EDI: What Managers Need to Know about the Revolution in Business Communications.* TAB Books, 1991, 312p., $32.95, ISBN: 0-8306-7724-0.

Baum, Michael S. and Perritt, Henry H. *Electronic Contracting, Publishing, and EDI Law,* Wiley Law Publications, 1991, 904p., $125.00, ISBN: 0-471-53135-9.

Gifkins, M. and Hitchock, D., Editors. *EDI Handbook: Trading in the 1990's.* Gower Pub Co, 1988, 337p., $120.00, ISBN: 0-86353-148-2.

Intl Research Dev. *EDI: Strategies and Stakes in the 1990s.* 1989, 265p., $2300.00, ISBN: 0-685-30584-8.

Kutten, L. J.; Reams, Bernard D.; and Strehler, Allen E. *Electronic Contracting Law: EDI and Business Transactions.* Clark Boardman Callaghan, 1991, $85.00, ISBN: 0-87632-825-7.

Management Advisory Pubns. *EDI Management. Control, Security and Audit,* 1992, 200p., $390.00, ISBN: 0-940706-27-X.

Marcella, Albert J., Jr. *EDI Audit and Control.* Artech House, 1992, 185p., $60.00, ISBN: 0-89006-610-8.

Metzgen, Fred. *Killing the Paper Dragon: Creating Business Advantage with EDI.* Butterworth-Heinemann, 1990, 200p., $29.95, ISBN: 0-434-91316-2.

Milbrandt, Ben. *EDI (Electronic Data Interchange) —Making Business More Efficient.* Data Interchange Standards Assn, 1987, 80p., $25.00, ISBN: 0-944952-00-3.

Sokol, Phyllis K. *EDI: The Competitive Edge.* McGraw Hill, 1989, 256p. $33.95, ISBN: 0-07-059511-9.

Walden, Ian, Editor. *EDI and the Law.* Gower Pub Co, 1989, 300p., $99.00, ISBN: 0-86353-170-9.

Wright. *The Law of Electronic Commerce: EDI, FAX, and E-Mail: Technology, Proof, and Liability.* Little, 1991, $95.00, ISBN: 0-316-95632-5.

PDES BOOK REFERENCES

Hunt, V. Daniel. *Enterprise Integration Sourcebook: The Integration of CALS, CE, TQM, PDES, RAMP and CIM.* Academic Press, 1991, 487p., $79.95, ISBN: 0-12-361777-4.

IGES BOOK REFERENCES

Mayer, Ralph. *CAD-CAM Standards: Is IGES the Answer?* Mgmt Roundtable, 1985, 300p., $425.00, ISBN: 0-932007-03-1.

Mayer, Ralph and Linden, Jonathan, Editors. *Making CAD-CAM Data Transfer Work: IGES and Other Solutions (a Hands-On Guide).* Management Roundtable, 1987, 250p., $295.00, ISBN: 0-932007-13-9.

Society of Automotive Engineers. *IGES (Initial Graphics Exchange Specification), Version 4.0.* 1988, $64.00, ISBN: 0-89883-694-8.

CALS BOOK REFERENCES

Hunt, V. Daniel. *Enterprise Integration Sourcebook: The Integration of CALS, CE, TQM, PDES, RAMP and CIM.* Academic Press, 1991, 487p., $79.95, ISBN: 0-12-361777-4.

Graph Comm Assn., *CALS Military Handbook.* 1988, 300p. $51.50, ISBN: 0-318-37101-4.

Dertouzos, Michael L.; Solow, Robert M.; and Lester, Richard K. *Made in America: Regaining the Productive Edge.* MIT Press. 1989, 248p., $22.50, ISBN: 0-262-04100-6.

GOVERNMENT PUBLICATIONS REFERENCES

Blake, M. W.; Chou, J. J.; Kerr, P. A.; and Thorp, S. A. *NASA-IGES Geometry Data Exchange Standard.* National Aeronautics and Space Administration, Moffett Field, CA, Ames Research Center, Apr 92, 6p, NTIS Number N92-24398/9.

Bridges, W. M. and Kaplan, B. J. *Implementation of Electronic Funds Transfer for Transportation Vendor Payment.* Logistics Management Inst.,

Bethesda, MD, Report No. LMI-PL005R2, Feb 92, 37p; NTIS Number AD-A251 446/1.

CALS: A Strategy for Change (English Version, 1/2 inch, VHS Format Video), National Inst. of Standards and Technology (CSL), Gaithersburg, MD, Dec 89, 1 VHS video, NTIS Number PB92-780931/XAB.

CALS in Print: 1990-1992. National Technical Information Service, Springfield, VA, May 92, 55p, NTIS Number PB92-962702/XAB.

CALS in Print: 1980-1989. National Technical Information Service, Springfield, VA, May 92, 53p, NTIS Number PB92-962701/XAB.

CALS Industry Steering Group. *CALS EXPO 90: Concurrent Engineering/PDES/STEP* (Video), Washington, DC, Dec 90, 1 VHS video, NTIS Number PB92-780360/XAB, Also available in set of 67 videos PB92-780006.

Clark, S. N. and Libes, D. *NIST PDES Toolkit: Technical Fundamentals*. National Inst. of Standards and Technology, Gaithersburg, MD, CALS Evaluation and Integration Office, Washington, DC, Report No. NISTIR-4815, 3 Apr 92, 33p; NTIS Number: PB92-187038/XAB; See also PB91-132159.

Carver, G. P. and Bloom, H. M. *Multi-Enterprise Concurrent Engineering through International Standards*. National Inst. of Standards and Technology, Gaithersburg, MD, Factory Automation Sys Div., Report No. NISTIR-4708, Oct 91, 27p; NTIS No PB92-123058; See also PB91-193367.

Christensen, M.; Farrell, J.; and Green, S. *IGES Efficiency Evaluation*. Lawrence Livermore National Lab., CA., Dept of Energy, Washington, DC, Report UCRL-JC-104493; CONF-901179-1, 1990 38p, NTIS No DE90015192.

Draft Handbook of Concurrent Engineering for Manufacturing, CALS/CE Industry Steering Group, Washington, DC, 1992, 81p, NTIS Number PB92-160456/XAB.

Drake, D. J., Logistics Management Inst., Bethesda, MD. *Electronic Data Interchange Opportunities in Defense Procurement*. Report No. LMI-DL203R2, May 92, 60p, NTIS Number AD-A252 664/8.

Frohman, H. L. *Guide to EDI Translation Software*. Logistics Management Inst., Bethesda, MD, Report No. LMI-PL005R1, 1991, 121p, NTIS Number AD-A233 053/8/XAB.

Henderson, M. M. and Lewis, A. P. *DOD Implementation Guidelines for Electronic Data Interchange (EDI), Volume 1*. Logistics Management Inst., Bethesda, MD., Report No. LMI-DL001-01R1, Dec 91, 134p, NTIS Number AD-A246 613/4/XAB.

Henderson, M. M. and Lewis, A. P. *DOD Implementation Guidelines for Electronic Data Interchange (EDI), Volume 2*. Logistics Management Inst., Bethesda, MD, Report No. LMI-DL001-01R1-VOL-2, Dec 91, 446p, NTIS Number AD-A246 614/2/XAB.

Information Resources Management Plan of the Federal Government Office of Management and Budget, Washington, DC. Office of Information and Regulatory Affairs, General Services Administration, Department of Commerce, Washington, DC, Report No. ISBN-0-16-036004-8, 20 Nov 91, 323p, NTIS Number PB92-147198/XAB.

Knoppers, J. V. T. *Results of the Work of the International Organization of Standardization and International Electrotechnical Commission on the "Open-EDI Conceptual Model" and Its Importance for EDI Developments*. Canaglobe International, Inc., Montreal (Quebec), 1991, 21p, NTIS Number PB92-169135/XAB.

Korzyk, A. D. *Architectural Guidelines for Multimedia and Hypermedia Data Interchange: CALS/CE and EC/EDI*. Naval Postgraduate School, Monterey, CA. Sep 91, 106p, NTIS Number AD-A246 201/8.

Linton, L.; Hall, D.; Hutchison, K.; Hoffman, D.; and Evanczuk, S. *First Principles of Concurrent Engineering: A Competitive Strategy for Electronic Product Development*. CALS Industry Steering Grp, Washington, DC, Report No CALS-TR-005, Sep 91, 181p, NTIS No PB92-102524.

Lindstrom, K.; Clark, P.; Fitzpatrick, J.; Klapper, L. *Logistics. Improving Supply Management: The CALS Connection*. Logistics Management Inst., Bethesda, MD, Report No. LMI-PL813R1; CALS/LMI/913 Feb 92, 121p NTIS Number AD-A252 214/2/XAB.

Luster, S. *Paperless Material Inspection and Receiving Report: Strategy to Streamline Acquisition and Reduce Paper*. Logistics Management Inst., Bethesda, MD, Report No. LMI-AF005R1, Mar 91,

160p, NTIS Number AD-A235 201/1/XAB.

Ross, E.; Lamb, J.; Fulton, R.; and Goclowski, J. *CALS 101: An Introductory Overview of CALS.* Intergraph Corp., Huntsville, AL, Army Materiel Development and Readiness Command, Alexandria, VA; Georgia Inst. of Tech., Atlanta, GA; CALS Industry Steering Group, Washington, DC, 1991, 121p, NTIS Number PB92-120393.

Sandia National Labs., Albuquerque, NM, *CAD/CAM Product Data Standards. IGES/PDES: The present and the future*. Department of Energy, Washington, DC, Rpt. No: SAND-90-0482C; CONF-9005151-1, 1990 5p, NTIS No: DE90008803.

Volpe, John A. *Production Definition Data Automation Plan.* National Transportation Systems Center, Cambridge, MA, Report No. DOD-VA056-90-5, Apr 90, 109p, NTIS Number PB92-120229/XAB; See also PB92-120211.

Zobrist, G. W. *Investigation of IGES for CADCAE Data Transfer*, Missouri University-Rolla Department of Computer Science, National Aeronautics and Space Administration, Washington, DC, Oct 89, 32p, NTIS No N90-16702/4.

Index

INDEX

A

B

C

D

E

F

P

Q

R

S

T

U